Civil Engineering Materials

Civil Engineering Materials

Editor

Sevilin Kartal

scitus
academics

Civil Engineering Materials
Edited by **Sevilin Kartal**

ISBN: 978-1-68117-591-1
Library of Congress Control Number: 2016946446

SCITUS Academics LLC,
616, Corporate Way, Suite 2, 4766,
Valley Cottage, NY 10989

www.scitusacademics.com

Notice

Printed in United States of America on Acid Free Paper

Preface

In addition to advanced materials, the modern civil engineer must handle standard construction materials. Traditional construction materials such as wood, steel asphalt and cement concrete from Portland are commonly used in many construction projects. Modern materials such as polymers and composites make progress in the construction industry. Research that is important on these materials has led to a better understanding of these materials and to the improvement of their durability and strength. The standard materials used today are much better than those of the past and new materials to meet the requirements of civil engineering applications are being developed in a unique way. The performance of materials in structures and their ability to combat various stresses are of key importance to a civil engineer.

Civil Engineering Materials focuses on demonstrating methods for obtaining, analyzing and using information rather than high data. It is designed to help civil engineering students understand the physical and structural characteristics of common building materials. It provides a comprehensive and detailed overview of traditional structural and civil infrastructure materials.

TABLE OF CONTENTS

1

CHAPTER

Bond Behaviour of Reinforcing bars in Cracked Concrete

Pieter Desnerck*, Janet M. Lees , Chris T. Morley

Department of Engineering, University of Cambridge, Trumpington Street, CB2 1PZ Cambridge, United Kingdom

HIGHLIGHTS

- A novel method to test bond in cracked reinforced concrete specimens is presented.
- Effect of number of cracks, orientation, confinement and cover is investigated.
- Results indicate a significant loss of bond strength for single cracked specimens.
- Concrete cover and confinement play a significant role.
- Crack orientation with respect to the rib pattern is of minor influence.

ABSTRACT

Due to the relatively low tensile strength of concrete, cracks are inherent in reinforced concrete structures. A common cause of cracking is the corrosion of internal steel reinforcement, a deterioration process that can affect the bond behaviour and anchorage capacity of reinforcing bars. Corrosion leads to a reduction of the reinforcing bar diameter, the formation of a weak layer of corrosion products around the bar and expansive forces on the surrounding concrete (that can lead to cracking). In the past, the impact of corrosion on bond has been investigated by means of accelerated corrosion tests. However, the more fundamental impact of cracking as distinct from corrosion products on the bond reduction is still not fully understood.

This study applies a novel test method to investigate the bond behaviour of reinforcing bars in cylindrical cracked reinforced concrete specimens. The influence of the number of cracks, crack orientation, confinement and concrete cover are investigated.

The results indicate a significant loss of bond strength for single cracked specimens. This reduction becomes as high as 65% for double cracked specimens in the absence of confinement. It is shown that the crack orientation with respect to the rib pattern is of minor influence, but the concrete cover and confinement play a significant role in the obtained bond characteristics.

Keywords: Bond strength; Corrosion; Pull-out strength

ABBREVIATIONS AND NOTATIONS

ϕ reinforcement bar diameter, mm

σ_s tensile stress in the reinforcing bar, N/mm^2

τ_M characteristic value of the bond strength, N/mm^2

τ_R ultimate bond strength, N/mm^2

BRF bond reduction factor, –

DEV standard deviation

f_c concrete compressive strength, N/mm^2

$f_{c,cub}$ concrete cube compressive strength, N/mm^2

$f_{ct,sp}$ concrete splitting tensile strength, N/mm^2

f_u ultimate tensile strength of a reinforcing bar, N/mm^2

f_y yield strength of a reinforcing bar, N/mm^2

F_s total tensile force in a reinforcing bar, N

k bond length expressed as 'times the bar diameter ϕ', –

l_d bond length, mm

s_u slip corresponding to the ultimate bond strength, mm

RBF retained bond factor, –

W crack width, mm

w_u ultimate crack width after bond tests, mm

1. INTRODUCTION

Concrete is an inhomogeneous material with a relatively low tensile strength. Therefore it is often used in combination with steel reinforcement so that the steel can resist tensile stresses after cracking. The load bearing capacity and serviceability performance of reinforced concrete structures depend highly on the interaction between reinforcing bars and the surrounding concrete [1]. Over time the bond can degrade due to deterioration of the reinforcement and/or the surrounding concrete. When cracks develop in regions surrounding the reinforcing bars, the force transfer is affected and this can lead to lower anchorage capacities or an altered bond behaviour.

Cracks are inherent in reinforced concrete structures and are caused by a number of different types of actions [2]. One of the most severe forms of cracking in hardened reinforced concrete structures is the result of the corrosion of steel reinforcing bars. Due to carbonation and chloride ingress a favourable environment is created for corrosion to initiate and corrosion products to be formed [2]. Fib bulletin No. 10 [1] categorises the effects of corrosion into 3 potential consequences:

- loss of reinforcing bar section,
- creation of a weak interfacial layer at the reinforcing bar/concrete interface,
- volumetric expansion of the reinforcing bar.

The last two phenomena have a direct influence on the bond of the reinforcing bar. On one hand, a weak interface layer of brittle corrosion products surrounding the reinforcing bar increases the relative displacement of the bar with respect to the concrete at certain load levels. On the other hand corrosion products are expansive in nature and tend to generate tensile stresses in the surrounding concrete. At low levels this expansion may be advantageous but once these stresses exceed the tensile capacity of the concrete, cracks around and along the reinforcing bars develop [1]. The reduction in confinement due to these cracks then leads to a progressive reduction in the bond strength. Hence, understanding of the link between corrosion rates, the induced cracking (and crack widths) and the reduction in bond capacity is essential.

To investigate crack formation due to corrosion, researchers [3], [4], [5], [6], [7], [8] and [9]have undertaken accelerated corrosion tests on reinforced concrete specimens where impressed currents of 0.003–10 mA/cm^2 were applied. Table 1 summarizes results found in literature on the corrosion levels to cause cracking of the concrete cover (defined by the appearance of visible cracks with a width of 0.03–0.05 mm at the outer surface of the specimens). The corrosion levels are expressed as a percentage of bar cross-sectional area loss (section loss expressed as a uniform metal loss around the circumference of the reinforcing bar) and corrosion penetration depth *x* radially into the bar. Although there is some scatter in the obtained values, all studies agree that a relatively low corrosion penetration depth, ranging from 0.008 to 0.130 mm, causes cracking.

Table 1. Overview of reported corrosion levels to cause concrete cover cracking.

Study	Bar diameter (mm)	Cover ratio (–)	Section loss (%)	Corrosion penetration *x* (mm)
Al-Sulaimani et al. [5]	10	7.50	5.00	0.130
	14	5.36	3.00	0.110
	20	3.75	2.00	0.100
Clark and Saifullah [8]	8	0.50–2.00	0.40–1.30	0.008–0.026
Andrade et al. [4]	16	1.25–1.88	0.40–0.50	0.015–0.020
Rodriguez et al. [3]	16	2.00–4.00	0.40–1.00	0.015–0.040
Clark and Saifullah [6]	8	1.00	0.30–0.55	0.006–0.011
Almusallam et al. [7]	12	5.00	4.00	0.120
Alonso et al. [9]	8–16	1.25–7.00	0.19–1.30	0.015–0.065

Rodriguez et al. [3] found that cracks were initiated for corrosion levels associated with bar radius reductions as low as 0.015 mm to 0.040 mm. But the cracking was delayed in specimens with an increased concrete cover and tensile strength. However Andrade et al.[4] reported a negligible effect of cover/bar ratio on cracking.

In previous studies, researchers have tried to link observed crack widths to the degree of corrosion in reinforcing bars. There is no agreed consensus on the exact relationship, but Vidal et al. [10] gathered data from beam tests with different bar diameters to deduce a linear relationship between crack widths w in mm and a loss of steel cross-sectional area of reinforcing bars:

$$w = K \cdot (A_{s0} - A_s) \tag{1}$$

where K is an empirical value of 0.0575 and $(A_{s0} - A_s)$ is the cross-sectional area loss of the reinforcing bar in mm^2.

A similar linear relationship, expressed as a function of the corrosion penetration depth x in mm, was proposed by Rodriguez et al. [3]:

$$w = 0.05 + \beta \cdot x \tag{2}$$

where β is a coefficient depending on the bar position in the element ($\beta = 0.0100$ for top bars and 0.0125 for bottom bars).

The bond reduction due to corrosion of reinforcing bars in concrete has also been the subject of investigation over recent decades [5], [7], [11], [12], [13], [14], [15] and [16]. Most studies apply accelerated corrosion techniques [17] to achieve a specific state of corrosion before the bond properties are assessed. The obtained results show a significant amount of scatter, but do agree on a bond reduction for significant levels of corrosion. Residual bond strengths of 15–110% are reported for cover ratios c/d ranging from 1.0 to 7.5 and higher, with values above 100% for very low corrosion levels (the initial formation of corrosion products can fill the pore structure around the bar and hence increase the bond strength slightly). The selected results shown in Figure 1 indicate the lack of a clear relationship between the cover ratio c/d and the retained bond factor (RBF). For more detailed information, reference is made to fib Bulletin No. 10 [1].

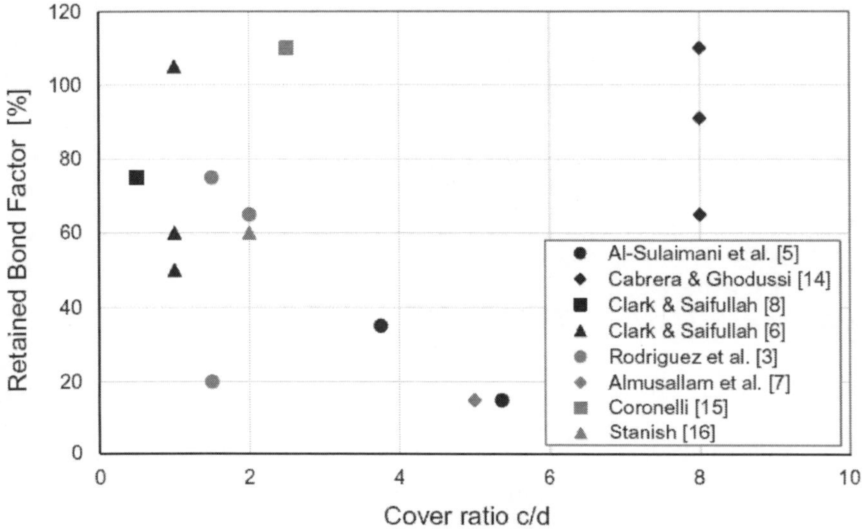

Figure 1. Relationship between cover ratio c/d and residual bond strength as reported in literature.

One issue with accelerated corrosion bond tests is that they study the combined effect of cracking and the formation of a soft layer of corrosion products around the reinforcement. While these effects co-exist in practice, it leads to difficulties in analysing the processes at a fundamental level. Furthermore the results cannot be applied to situations in which cracks are present in the absence of corrosion (e.g. plastic shrinkage cracks in congested reinforcement lay-outs).

This study aims to quantify the bond strength of reinforcing bars in cracked concrete. Rather than performing accelerated corrosion tests, it focuses on the more fundamental effect of the cracking itself (in absence of corrosion products). To achieve this aim, a novel test method is developed.

2. CRACK GENERATION

When assessing existing structures, engineers often observe severe cracking due to corrosion. Depending on the location of the reinforcing bar within the structure, the crack orientation and extent can differ, which in turn can influence the bond properties. Some common corrosion induced crack patterns are shown in Figure 2.

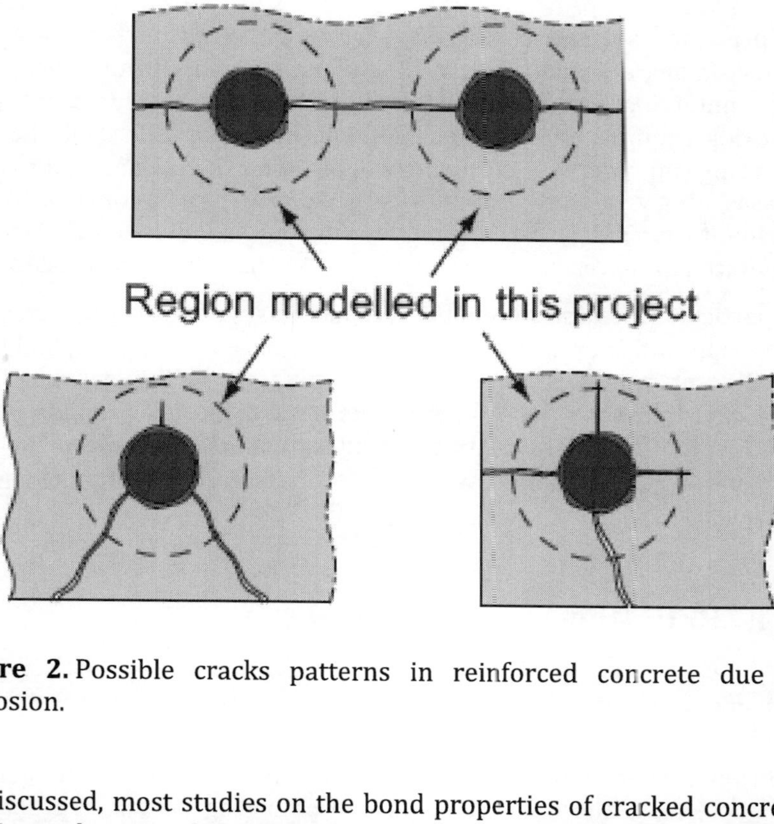

Figure 2. Possible cracks patterns in reinforced concrete due to corrosion.

As discussed, most studies on the bond properties of cracked concrete specimens have used accelerated corrosion techniques. However, these techniques have the tendency to lead to less controllable crack patterns and a wide range of obtained crack widths. Gambarova et al. [18] studied the bond of reinforcing bars in pre-cracked specimens. The cracks were formed by applying a transverse tensile force on a specially designed concrete element (steel shoulders with shear pins were cast into the specimen to allow for the application of the force). The specimens were relatively large and the test set-up consisted of external confinement provided by a 'confinement cell'. In research projects focussing on diffusion of contaminants in concrete, cracks are often introduced in one of the following ways: artificially by using thin inserts (e.g. metal sheet) [19], 3-point bending tests, wedge splitting tests [20], mechanical expansive cores [21] or controlled splitting tests [22]. As with respect to bond tests, use of inserts (eliminating the aggregate interlock between the crack faces), wedge shaped specimens or an expansive core are not easy to implement and are not necessarily representative of cracked concrete.

In this study the principles of a controlled split tensile test are applied to pre-crack specimens. In this way rough crack surfaces are then formed along a predefined cracking plane running through the axis of the reinforcing bar. To avoid brittle failure and provide some post-cracking confinement, the specimen can be cast within a plastic ring. This ring can remain in place during both the pre-cracking and pull-out phases. Singly reinforced cylindrical specimens are well suited to split tensile tests. With this technique, single or double cracks can be generated to represent observed crack patterns in existing structures.

Of particular interest are the bond reductions during the early stages of corrosion and thus the onset of cracking. Based on Eqs. (1) and (2), a 1% loss of steel section corresponds to crack widths of about 0.03–0.04 mm, therefore the splitting tests seek to achieve similar values. These crack widths are relatively small compared to the allowable crack widths of 0.20–0.30 mm according to reinforced concrete design guidance [23].

3. EXPERIMENTAL PROGRAM

3.1. Experimental specimens and overview of test method

In order to represent common crack patterns (as illustrated in Figure 2) and determine the bond strength of reinforcing bars in cracked, the proposed test method includes one or two pre-cracking phases followed by standard bond strength "pull-out tests" [24].

In the pre-cracking phase(s) the specimens are subjected to a split cylinder test. Two line loads are applied on the specimen on opposite sides and along the axis of the concrete cylinder until first cracking of the concrete, at which point the specimen is unloaded. For double cracked specimens, the specimen is rotated 90° after the first pre-cracking experiment and the same procedure is repeated to induce the second crack. Pull-out tests are then conducted on the cracked specimens to determine the influence of the cracks on the bond behaviour. The test procedure is shown schematically in Figure 3.

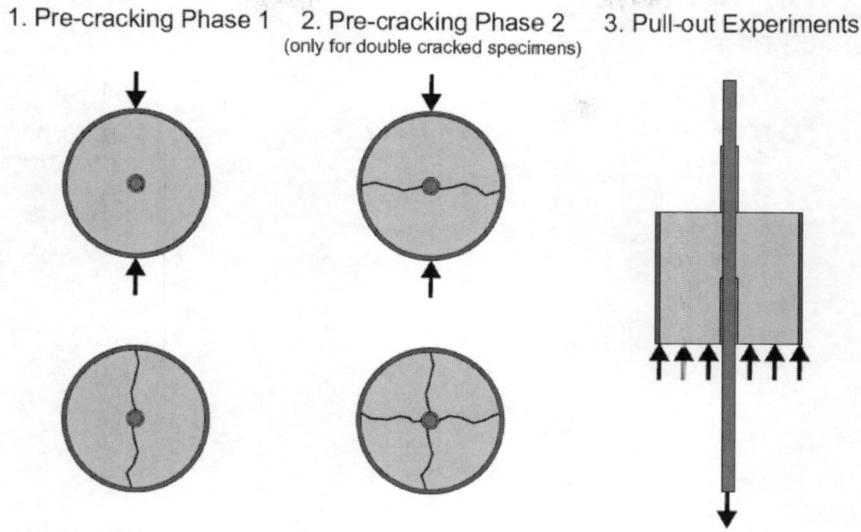

Figure 3. Test procedure.

Cylindrical specimens with a diameter of either 107 mm or 60 mm, and a height of 100 mm are used. A reinforcing bar with a diameter of 10 mm is centrally placed in the specimen, resulting in a cover-to-diameter ratio of 4.8 or 2.5. The prescribed bond length for pull-out tests according to RILEM recommendations [24] is 10 times the bar diameterϕ. However, as shown in other studies [25], [26] and [27], this can lead to yielding and in some cases the rupture of the reinforcement bar before reaching the ultimate bond strength. Therefore the specimens are cast with a bond length of 5 times the bar diameter. The bond length itself is reduced by providing a sleeve over the bar in the lower 50 mm (Figure 4). The specimens are cast in a plastic cylindrical mould which is used as confinement for the specimen during the pre-cracking phases and in many cases also remains in place for the pull-out tests (the plastic confinement is not removed at the time of demoulding).

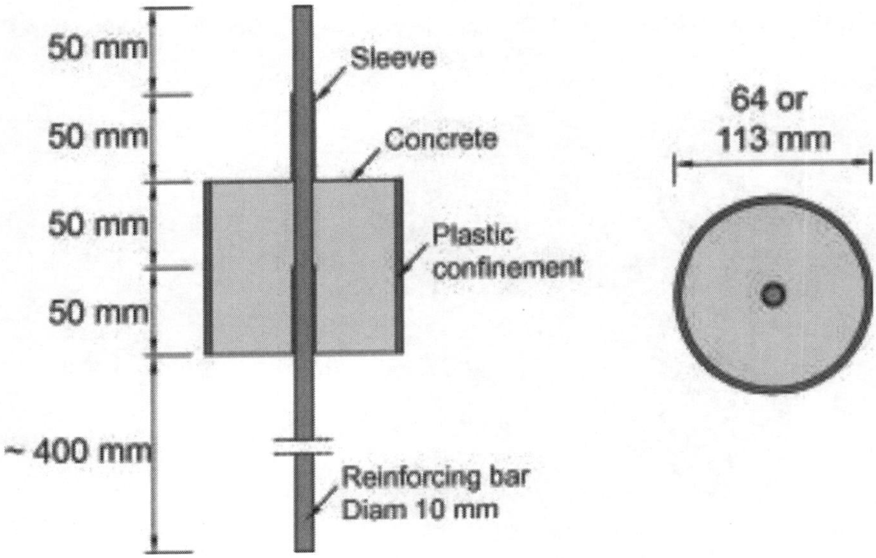

Figure 4. Specimen geometry.

In order to be able to install measuring devices on the passive side (free end of the bar), the reinforcing bar extends 100 mm beyond the concrete surface. At the active end the reinforcing bar extends 400 mm beyond the concrete to provide room for sensors and provide a sufficient length to grip the bar in the clamps of the testing machine.

3.2. Main parameters

The following parameters were selected for investigation: single or double pre-cracks (number of cracks), the crack orientation relative to the reinforcing bar rib pattern, the confinement and the concrete cover.

As reinforced concrete structures are currently primarily constructed with ribbed reinforcing bars, the crack itself can cross the rib pattern of the bar at different angles (see Figure 5). Cracks running along the longitudinal ribs, leaving the concrete around the transverse rib intact, might have a less detrimental impact on the bond compared to cracks that hit the transverse ribs. Whether this crack angle influences the bond behaviour is currently unknown.

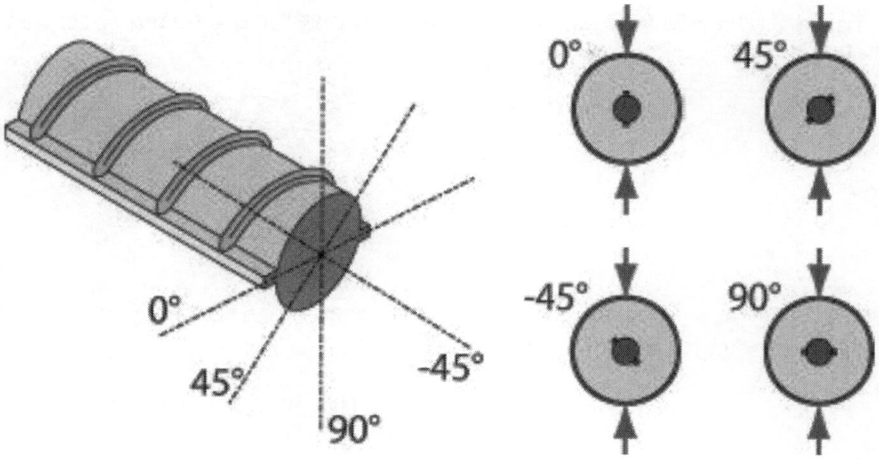

Figure 5. Applied crack orientations with respect to the reinforcing bars rib pattern.

Crack widths are often governed by the amount of transverse reinforcement. For longitudinal cracks running along the longitudinal reinforcing bars, the transverse reinforcement provides a form of confinement transferring forces from one crack face to another and reducing the crack width. As more confinement is provided, the hoop stresses induced during pull-out of a longitudinal reinforcing bar might be counteracted and in this way provide sufficient resistance for the anchorage force to build up [28]. However, if the confinement is limited these hoop stresses can lead to the cracks opening up and a decrease in the bond properties.

Similar to the confinement provided by transverse reinforcement, the concrete cover can also provide confinement. Increasing the concrete cover leads to a bigger influence zone around the reinforcing bar that can resist hoop stresses and thereby increase the bond strength of the bar.

3.3. Test series

A total of four series of experiments are undertaken. Each series consists of 4 sets of 5 identical specimens. In this way five individual results are obtained for each parameter combination. The entire test matrix for the total of 80 tests carried out is shown in Table 2.

Table 2. Overview of parameter combinations for the tested specimen series.

Series	Set	Confinement		Diameter (mm)		Single crack angle			Double crack angle		
		No	Yes	60	107	0	45	90	0 and 90	−45 and 45	90 and 0
1	A		X		X						
	B		X		X	X					
	C		X		X		X				
	D		X		X			X			
2	A		X		X						
	B		X		X				X		
	C		X		X					X	
	D		X		X						X
3	A	X			X						
	B	X			X	X					
	C	X			X			X			
	D	X			X				X		
4	A		X	X							
	B		X*	X*	X*	X					
	C		X	X		X					
	D		X	X					X		

* Pre-cracked specimen of 60 mm cast in a 107 mm specimen

In Series 1 the angle of a single crack with respect to the rib pattern of the bar is varied. Different angles are achieved through different orientations of the specimens relative to the applied split tensile load (Figure 5). In addition to a reference uncracked set, a single cracked set with crack angle of 0°, 45° and 90° is tested. The crack angle itself is measured with respect to the plane running through the bar axis and the longitudinal ribs (see Figure 5).

Series 2 focusses on double cracked specimens where the crack orientation relative to a reference plane through the bar axis and longitudinal ribs varies. A reference set, a set with cracks orientated at 0° and 90° relative to the reference plane, and a set with −45° combined with 45° cracks are tested. As the cracks are formed by applying a splitting test, the crack faces of the first crack are compressed during the second splitting phase. To investigate the potential influence of this sequence, a fourth set of specimens was included. Instead of pre-cracking in the 0° direction (as in set 2-B), in set 2-D the first crack was applied at 90° followed by the one at 0°. In this way, the predominant crack (with the biggest crack width) in set 2-D was the one perpendicular to the longitudinal ribs, whereas in set 2-B it was the one parallel to the longitudinal ribs.

Series 3 consisted of specimens where the plastic tube confinement was removed after the cracking phase. The different parameter combinations tested without confinement are uncracked reference specimens, single cracked specimens with cracks orientated at either 0° or 90° to the reference plane and a set of double cracked specimens with a crack at 0° combined with a crack at 90°.

In the last series, Series 4, the concrete cover was reduced by casting the specimens in a cylindrical tube with an inner diameter of only 60 mm, reducing the original cover by 23 mm and resulting in a cover-to-diameter ratio of 2.5. An uncracked reference set, a set of specimens with single crack (crack angle 0°) and a set of double cracked specimens (0° combined with 90°) are tested. The final set in the series consisted of specimens of diameter 60 mm single cracked at 0° after which the plastic confinement was removed. The specimen was then placed in the centre of a 107 mm diameter plastic tube and concrete was cast in the annulus between the specimen and the tube. As a result a 23 mm thick concrete ring was formed around the cracked specimen. This scenario simulates a situation in which the crack starts to grow around the rebar but does not yet extend to the outer surface.

3.4. Materials

A concrete with a 28-day cube compressive strength of about 30 MPa is selected for the tests. The water-to-cement ratio of the mixture is 0.6 and the sand-to-aggregate ratio is 0.45. The concrete is made with an early-strength Portland cement with limestone, type CEM II/A LL 32.5 R [29]. The aggregates are a natural sand with a maximum grain size of 4 mm and specific gravity of 2.65 and a gravel with maximum grain size of 10 mm with specific gravity of 2.65. The mix proportions are summarized in Table 3.

Table 3. Concrete composition.

Constituent	Density (kg/m^3)	Quantity (kg/m^3)
Water	1000	180
Sand	2625	835
Coarse aggregate	2625	1015
CEM II/A LL 32.5 R	3100	300

The mixtures are prepared in batches of 40 L from which approximately 20 L are used for the pull-out specimens and the other 20 L for standard concrete control specimens. After 1 min mixing of the dry materials, water is added and mixing continues for 3 min.

After casting, the specimens are stored in standard lab conditions (temperature of 22 ± 2 °C and relative humidity of 70 ± 5%). After 24 h the control specimens are demoulded and stored under water at a temperature of 20 ± 2 °C. The pull-out specimens are taken from the formwork (without removing the plastic confinement rings), wrapped in plastic and stored in standard lab conditions.

The compressive and tensile strengths of the concrete are determined at 21 days (time of pre-cracking) and 28 days (time of bond tests). For the determination of the compressive strength, cubes with sides of 100 mm ($f_{c,cub}$) and cylinders with a height of 200 mm and a diameter of 100 mm (f_c) are used. The splitting tensile strength $f_{ct,sp}$ is measured from cylinders with a diameter of 100 mm and a height of 200 mm. The mean results for the four series are summarized in Table 4.

Table 4. Concrete properties.

Series	$f_{c,cub}$ (MPa)		f_c (MPa)		$f_{ct,sp}$ (MPa)
	21 days	28 days	21 days	28 days	21 days
1	29.5	31.3	22.8	23.3	2.4
2	29.0	30.5	20.0	20.0	2.6
3	25.3	27.6	20.2	22.5	2.5
4	28.4	29.0	23.4	23.8	2.7

For the test program, the nominal steel bar diameter is fixed at 10 mm (actual measured value was 9.92 mm). The yield stress f_y and ultimate tensile strength f_u of the high-strength hot-rolled reinforcing bars, are measured in the laboratory as 520 MPa and 606 MPa respectively. The stress–strain diagram is given in Figure 6(a).

All reinforcing bars have the same rib pattern (see Figure 7) which consist of two longitudinal ribs at opposite sides of the bars and two series of transverse ribs. The rib pattern is measured in the laboratory using a laser measuring device. The transverse ribs are parallel, at 48° to the longitudinal ribs, and of maximum rib height 0.75 mm.

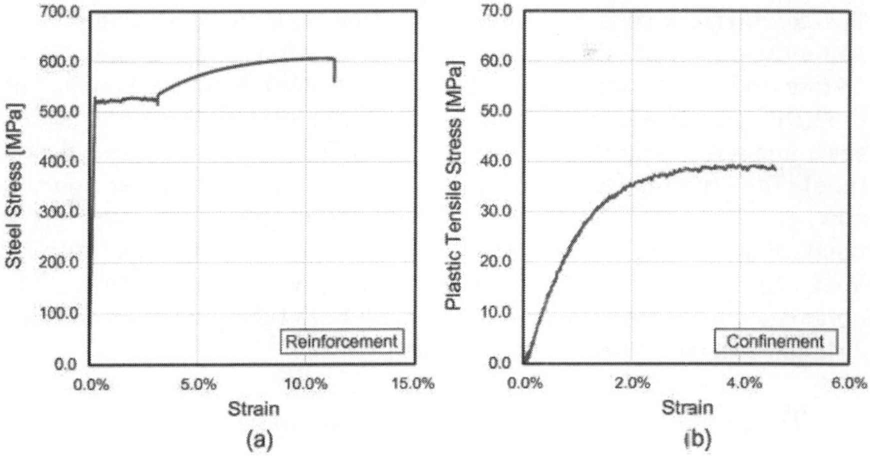

Figure 6. Stress–strain curves of (a) steel reinforcement bar (diam. 10 mm) and (b) HDPE plastic confinement.

Figure 7. Rib pattern of reinforcing bars diameter 10 mm.

The confinement consisted of a plastic tube with an outer diameter of 113 mm with a wall thickness of 3.2 mm or a tube with a diameter of 64 mm and a wall thickness of 2.1 mm. The stress–strain behaviour of the HDPE plastic was measured in a tensile test set-up, using a dog-bone specimen with a width of 20 mm (the specimen was cut from the plastic tube). The stress–strain response of HDPE is highly dependent on the applied strain rate[30]. As the strains in the confining ring during the bond pull-out test are being built up slowly, a low strain rate of 5×10^{-5} was applied during the material characterisation tests. The obtained results are shown in Figure 6(b). Both tube diameters showed a similar behaviour under tensile loads.

3.5. Test procedure and instrumentation

At an age of 21 days the specimens are subjected to a split cylinder test according to EN12390-6 [31] (Figure 8). The test is performed in a rigid load frame with a capacity of 150 kN. A constant deformation rate of 0.4 mm/s is applied until first cracking, at which point the specimen is unloaded. During the splitting test, the applied force as well as the strain on the surface of the plastic tube is recorded. The tube confining strain is measured with strain gauges on opposite sides of the specimens at mid-height.

Figure 8. Instrumentation and test set-up for the pre-cracking phase.

Figure 9. Instrumentation and test set-up for the pull-out experiments.

The obtained crack width is measured using a crack microscope with an accuracy of 0.01 mm at both sides of the reinforcement bar at a distance of 30 mm from the rebar in case of 107 mm specimens and at 20 mm for specimens with a diameter of 60 mm.

The pull-out tests are performed at 28 days in a 150 kN closed loop tensile load frame. The concrete is supported by a steel plate during the test while a tensile force is applied on the reinforcing bar. During the pull-out tests, the specimens are loaded at a constant displacement rate of 0.04 mm/s until a total relative displacement between the tested reinforcing bar and the surrounding concrete of at least 12.5 mm is reached.

The slip of the bar, at the passive as well as the active side, is recorded using 2 sets of 2 linear variable differential transducers (LVDT). These LVDT's are secured to the bar by means of a steel collar mounted to the reinforcing bar (Figure 9).

After unloading of the specimens the crack widths are measured again.

4. RESULTS AND DISCUSSION

4.1. Pre-cracking of the specimens

For the pre-cracking stage, the measured applied forces at first cracking induce splitting stresses, calculated based on the principles of standard splitting tensile tests, of around 2.3 MPa (DEV = 0.2 MPa). These splitting strengths are, due to the presence of the reinforcing bar in the splitting plane, slightly lower than the tensile splitting strengths measured on standard cylindrical concrete specimens.

Figure 10 shows one of the measured splitting stress – confinement strain responses for a Series 1 – set B, C and D specimen (single crack oriented at either 0°, 45° or 90° relative to the longitudinal rib). The responses of the other specimens are comparable. As the splitting stress builds up, small compressive stresses are measured in the plastic confinement up until the point where cracking occurs. Thereafter the circumferential strains increase abruptly and become tensile. The final strain in the confinement after cracking is in the 0.04–0.06% range which corresponds (based on the measured stress–strain diagram) to confinement stresses of 1.8–2.3 MPa in the plastic confinement ring.

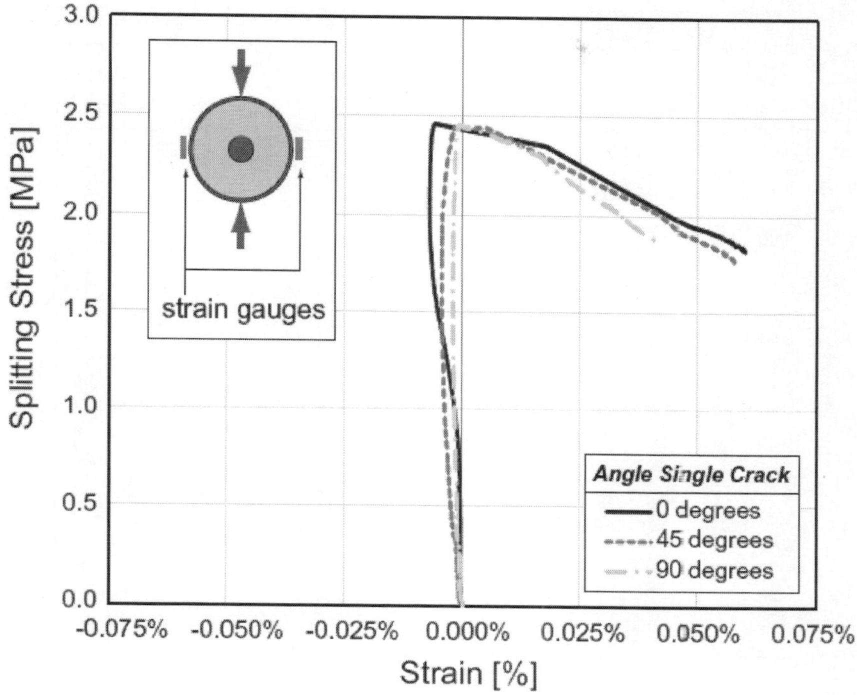

Figure 10. Confinement strains during pre-cracking tests for single cracked specimens.

The crack widths *w* after splitting for the pre-cracked specimens are shown in Figure 11 for the different series and sets. The obtained values are around 0.03 mm but there is significant scatter. Although cracks were visible for all specimens, the 95% confidence intervals of some sets (indicated by the error bars in the figure) included 0.00 mm crack widths. This is due to scatter values of around 0.02 mm for these specific sets. For the double cracked specimens (Series 2) the average of the two (phase 1 and phase 2) obtained crack widths is plotted. Figure 12 shows the crack patterns for a typical single and double cracked specimen as well as for a single cracked specimen that is then re-cast in a concrete ring. All crack patterns obtained for specimens within the same parameter combination were comparable. No anomalies were detected.

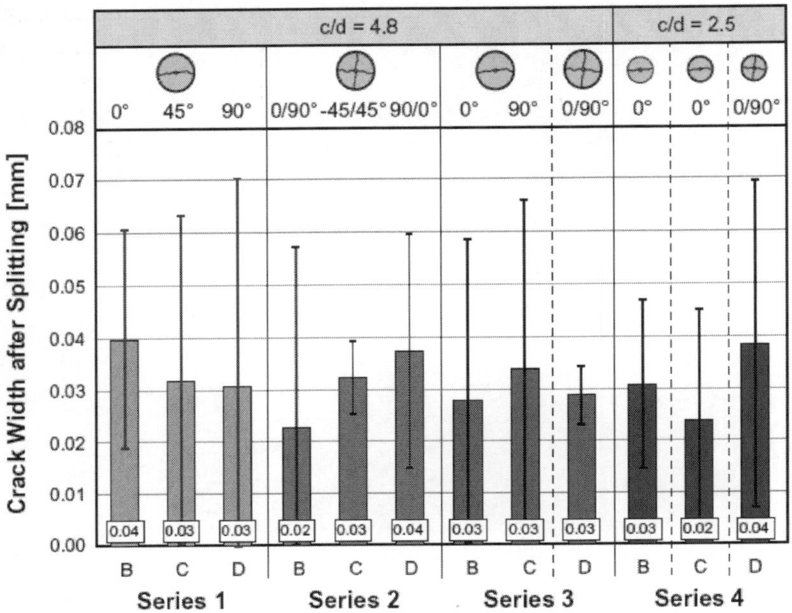

Figure 11. Crack widths *w* after pre-cracking phase.

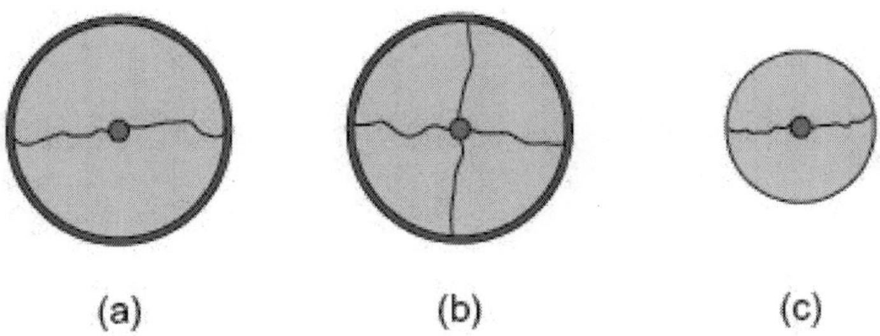

Figure 12. Observed crack pattern for a typical (a) Series 1B, (b) Series 2B and (c) Series 4B specimen.

5. BOND STRENGTH

From the pull-out experimental results, values of the bond stress can be derived. Assuming a uniform stress distribution along the short bond length, the force F_s in the reinforcing bar is transferred to the concrete over the embedment length l_d resulting in mean bond stresses τ_d of:

$$\tau_d = \frac{F_s}{l_d \cdot \pi \cdot \phi} = \frac{\sigma_s}{4 \cdot k}$$

where σ_s the tensile stress in the reinforcing bar and l_d as a parameter equal to $k$multiplied by the bar diameter ϕ ($k{\cdot}\phi$).

Two values are of particular interest: the ultimate bond strength τ_R and the often used characteristic bond strength τ_M as defined by RILEM [24]. The ultimate bond strength is defined as the bond stress corresponding to the ultimate load recorded during testing. The characteristic bond strength τ_M is calculated as the mean value of the bond stresses corresponding to a slip of 0.01 mm, 0.10 mm and 1.00 mm.

The slip at the ultimate bond strength s_u is of interest with respect to the ductility of the bond failure.

Table 5 summarizes the obtained results for the different series and sets. The highest ultimate bond strengths are recorded for the reference uncracked specimens, as was to be expected. The three sets of 107 mm diameter reference specimens (c/d = 4.8) all gave ultimate bond strengths of about 19.2 MPa (DEV = 1.4 MPa). The bond strength of the bars tested with a smaller cover (c/d = 2.5 for the 60 mm diameter specimens), turned out to be 14.3 MPa (DEV = 0.8 MPa) as the dominant failure method in this case was a splitting failure (due to the limited reinforcement cover). The ultimate bond strengths of the pre-cracked specimens were all lower than the equivalent reference specimens. The same observations can be made for the characteristic bond strength.

In the analysis hereafter the results of the bond stresses and strengths will be expressed in terms of a retained bond factor (RBF). This factor is defined as the ratio of the actual strength of a specific set to the mean bond strength of the uncracked reference specimens (mean value of all results from sets 1-A, 2-A and 3-A). In this way direct comparisons of the test results within and between series can be made. The bond reduction factor (BRF) can be obtained from the retained bond factor as follows:

$$BRF = 1 - RBF$$

Table 5. Experimental results of pull-out test.

Series	Set		τ_R (MPa)		τ_M (MPa)		s_u (mm)		w_u (mm)	
			Mean	DEV	Mean	DEV	Mean	DEV	Mean	DEV
1	A	Reference	20.2	1.3	11.0	0.9	0.95	0.09	0.00	0.00
	B	Single	11.4	0.4	5.5	0.3	0.56	0.09	0.16	0.06
$c/d = 4.8$	C		11.5	0.5	6.1	0.3	0.58	0.11	0.19	0.06
Confined	D		11.1	0.6	5.6	0.2	0.57	0.10	0.18	0.08
2	A	Reference	19.6	1.0	11.1	1.4	0.89	0.09	0.00	0.00
	B	Double	8.7	0.6	4.0	0.4	0.65	0.13	0.08	0.04
$c/d = 4.8$	C		9.5	0.6	4.3	0.4	0.63	0.12	0.07	0.01
Confined	D		9.0	0.8	4.4	0.5	0.69	0.13	0.07	0.03
3	A	Reference	17.9	0.9	10.3	1.0	0.92	0.14	0.00	0.00
	B	Single	8.0	0.7	2.9	0.4	0.41	0.12	2.02	0.70
$c/d = 4.8$	C		8.1	1.0	3.4	1.0	0.54	0.12	1.62	0.49
Unconfined	D	Double	6.4	1.2	2.3	0.6	0.38	0.05	0.89	0.37
4	A	Reference	14.3	0.8	8.2	1.1	0.45	0.23	0.12	0.03
	B*	Single	16.1	0.6	9.3	1.7	0.43	0.24	0.36	0.59
$c/d = 2.5$	C	Single	10.6	0.9	7.1	1.5	0.31	0.12	0.13	0.01
Confined	D	Double	8.4	0.9	4.5	0.6	0.61	0.20	0.08	0.02

*Re-cast = specimen with $c/d = 2.5$ pre-cracked and re-casted to achieve $c/d = 4.8$.

5.1. Crack orientation

The influence of the crack orientation in single cracked specimens was investigated in Series 1. Three different crack angles were considered, 0°, 45° and 90°, with respect to the longitudinal ribs. The results clearly show that the obtained bond strengths are all lower than the reference bond strength (Figure 13). A *t*-test with a significance level of 95% confirmed this difference to be significant. The differences between the studied crack angles is small and statistically insignificant. Therefore it can be concluded that the orientation of the crack relative to the longitudinal rib has little to no effect on the bond strength. The presence of a single crack however results in a significant reduction in the ultimate bond strength of 44% (RBF = 0.56) for the tested specimens. Those values are slightly higher than the retained bond factors reported earlier (see Figure 1) for specimens with c/d equal to about 4–5. However, when a wider range of concrete covers is considered, a value for the retained bond factor of 0.56 is well within the reported range of 0.15–1.10.

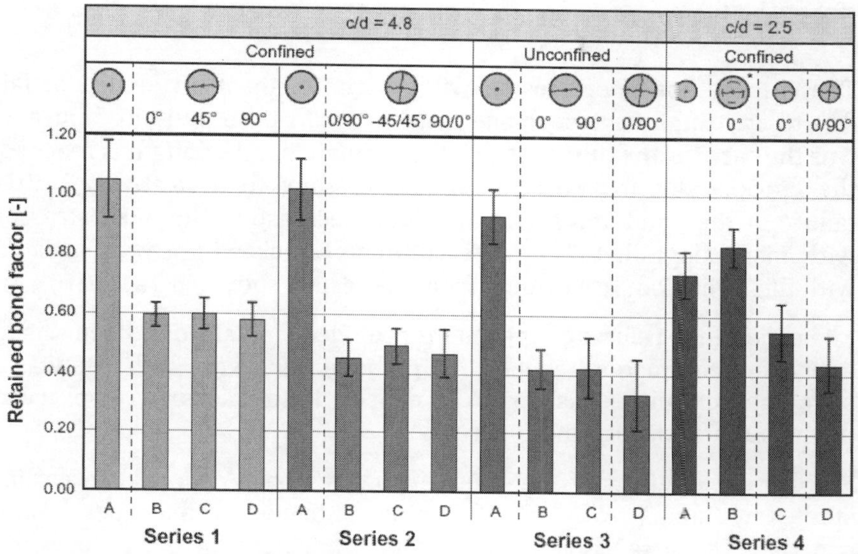

* Re-cast = specimen with c/d = 2.5 pre-cracked and re-casted to achieve c/d = 4.8

Figure 13. Bond reduction with respect to uncracked reference specimens.

5.2. Multiple cracks

As was the case for the single cracked specimen, the orientation relative to the longitudinal rib is insignificant, as the 0–90° and −45–45° double cracked specimens gave similar results. In both series the cracks led to ultimate bond strengths that are 54% lower than the uncracked specimens (RBF = 0.46).

The similarity of the set 2-B and 2-D results (double crack with angles 0° and 90°, and 90° and 0°, resp.) suggests that the order in which the cracks were formed also had little effect on the bond performance, as was to be expected from the results of the crack orientation tests. In comparison to the single cracked specimens, the double cracks tend to provide an additional reduction in bond strength of about 10%.

5.3. Confinement

The cracked specimens tested without confinement all tended to fail due to splitting of the specimen rather than a pure pull-out failure as was the case for the other sets. As the ultimate bond strength is reached, the crack width increases drastically leading to a relatively brittle failure of the bond and hence a sudden decrease in the bond stresses with increasing slips. The larger ultimate crack widths w_u associated with this behaviour are evident from the results shown in Table 5.

The unconfined reference specimens provided similar bond strengths to equivalent confined specimens. As the bond failure for both sets was a pure pull-out failure this was to be expected. No cracks were observed on the outer surface of the concrete.

The beneficial effect of the confinement has been reported in earlier studies by Gambarova et al. [18] and is confirmed in this experimental program. The absence of any external confinement further reduces the bond capacity of the 10 mm reinforcing bars. For single cracked specimens, an additional reduction of 11% is measured whereas for the double cracked specimens this reduction is 20%.

5.4. Concrete cover

As the concrete cover is reduced in Series 4 to a cover-to-diameter ratio of 2.5, the failure mode of the reference specimens shifts from a pull-out failure to a splitting failure, hence a reduction in the ultimate bond strength from 19.2 to 14.2 MPa. The single cracked and double cracked specimens show a reduction relative to the reference of 25% (RBF = 0.75) and 41% (RBF = 0.59) respectively, which is slightly less than for the counterparts with a 4.8 cover-to-diameter ratio. This difference can be attributed to the changed failure mode.

For the specimens which originally have a cover-to-diameter ratio of 2.5 on pre-cracking, but are recast to give a cover-to-diameter ratio of 4.8, an increase in bond strength is measured. Compared to the original uncracked specimens with a c/d-ratio of 2.5, the increase is 21%. The concrete ring with a thickness of 23 mm is sufficient to carry the hoop stresses built up during the pull-out test and hence prevent a splitting failure. Failure in these specimens occurs due to pull-out. The cracks in the inner cylinder do not extend to the surface of the uncracked concrete. Comparing these specimens to the original specimens with a cover-to-diameter ratio of 4.8, shows that the re-cast specimens are stronger than the single cracked specimens. This suggests that the bond strength is less affected by partial cracking than in the case where the cracks extend to the outer surface of the concrete. However, the bond strength is lower than measured for uncracked reference specimens with $c/d = 4.8$.

5.5. Implications for design

Current guidelines and standards take into account the effects of micro-cracking, small voids underneath reinforcing bars, potential light forms of corrosion, etc. by applying a relatively high safety factor on design bond strengths. The design implications of the experimental observations of a reduced bond strength due to cracking have not yet been fully identified. More work is required to relate the values obtained in this study to design values, as those might already – perhaps without explicitly realising it – allow partially for the effect of cracks along the bar.

6. BOND STRESS – SLIP RELATIONSHIP

During the pull-out tests, the relative displacement between the reinforcing bar and the surrounding concrete is measured so the bond stress–slip relationship can be plotted.

For a given set of 5 specimens, the results show a high repeatability as illustrated in Figure 14 for Series 1-B (single cracked specimens with cracks at 0°). The measurements on the passive side gave similar slip values to those measured on the active side after correction for the linear elastic deformation of the reinforcing bar.

As stated previously, the crack orientation does not significantly influence the obtained characteristic or ultimate bond strength. A comparison of the bond stress – slip diagrams obtained for 0°, 45° or 90° crack angles (Figure 15), confirms that the crack orientation relative to the reinforcing bars longitudinal rib has little to no influence on the bond response.

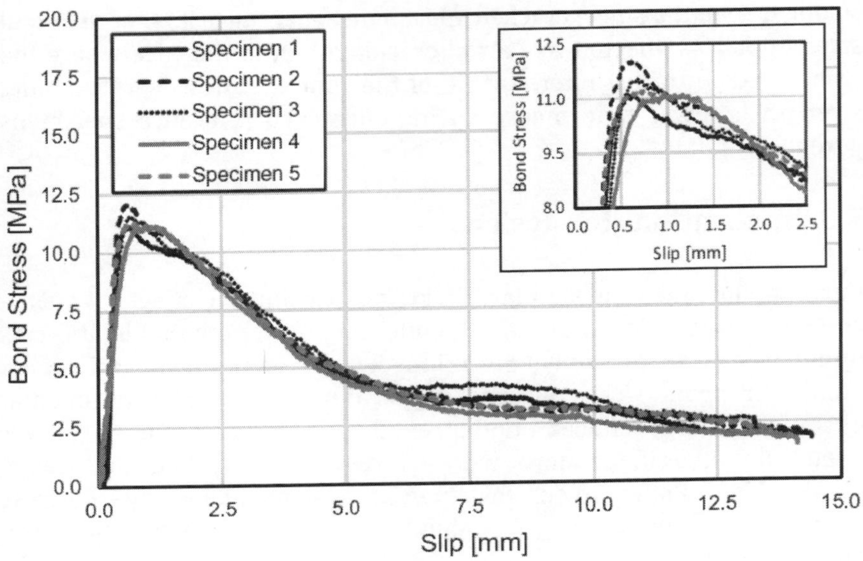

Figure 14. Bond stress – slip diagram for single cracked specimens with cracks angles of 0° (Series 1 B).

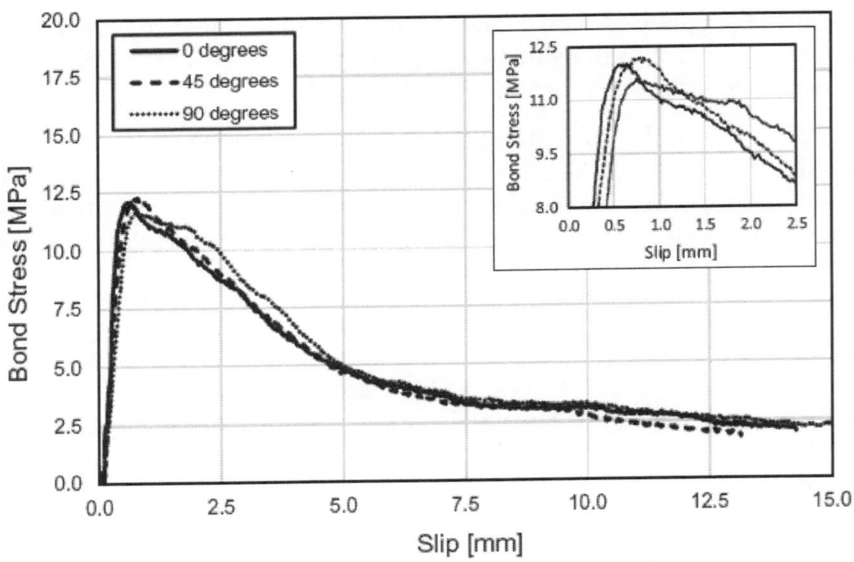

Figure 15. Influence of crack angle on the bond stress – slip diagram for single cracked specimens.

A plot of the bond stress–slip response of a typical uncracked, single cracked and double cracked specimen shows that the slope of the initial ascending branch decreases with an increasing number of cracks (Figure 16), indicating a decrease in stiffness. As the bond stresses increase, the slip seems to increase in an almost linear way. As soon as the ultimate bond strength is reached however, a steeper descending curve is measured for uncracked specimens when compared to single or double cracked ones. Thereafter, the bond stresses recorded for higher slip values decrease with an increasing number of cracks, indicating not only the ultimate strength, but also the post-failure behaviour is influenced by the cracks.

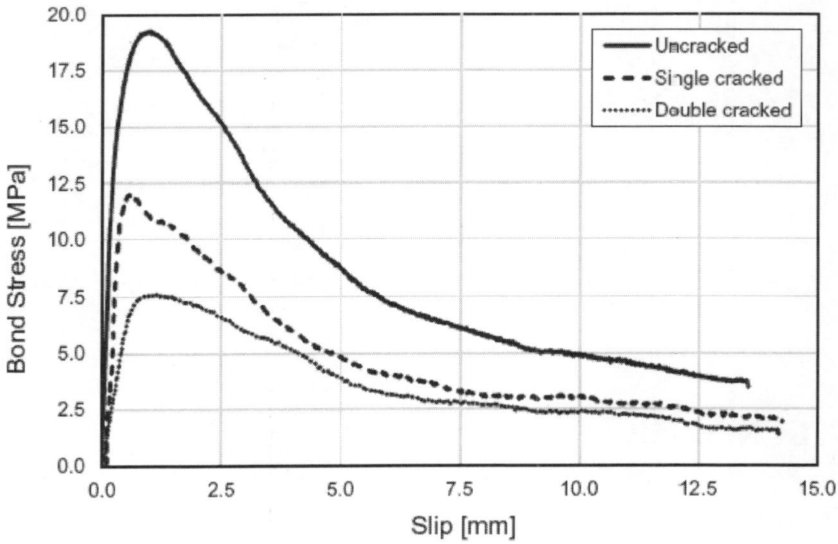

Figure 16. Influence of the number of cracks on the bond stress – slip diagram.

The bond stress – slip behaviour of the uncracked reference specimens is not influenced by the plastic tube for the higher concrete cover ratio c/d of 4.8 (Figure 17). In contrast, for single cracked specimens, the confinement has a significant impact on the bond behaviour. With respect to the recorded slips, a sudden drop in bond stress is noticed for small slip increments beyond the ultimate bond strength. Hence a more sudden bond degradation is noted due to opening up of the cracks as shown by the measured w_u values when there is no confinement. Reducing the concrete cover in a single cracked confined specimen has an effect, but the overall shape is comparable and the post-peak behaviour beyond a slip of approximately 5 mm is similar.

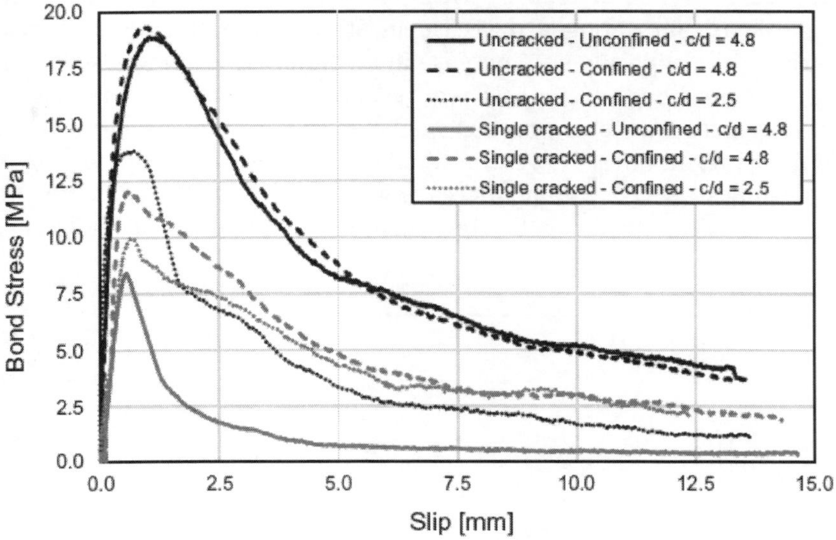

Figure 17. Influence of confinement on the bond stress – slip diagram.

7. CONCLUSION

Reinforcement corrosion can result in a severe form of concrete cracking. As cracks due to corrosion run along the concrete – rebar interface, the force transfer between the two materials, and more specifically the bond behaviour, is influenced.

To study the effect of the crack angle (relative to a plane running through the axis of the bar and the longitudinal ribs), extent of cracking, confinement and cover depth on the bond properties of reinforcing bars embedded in concrete, an experimental test program was conducted. Cylindrical specimens with a central 10 mm diameter reinforcing bar and a compressive strength of 25 MPa were pre-cracked by applying split tensile forces. Pull-out tests were then carried out on the cracked specimens to determine the bond properties. The results of the tests indicated that:

(1) The presence of cracks, even with minor crack widths of 0.03–0.04 mm, result in a significant reduction of the bond strength. For specimen with a single crack the reduction was on average 44% and for double cracked specimens the reduction was 54%. The measured values for single and double cracked specimens are within the relatively wide range of values reported by other researchers in the past.

(2) The crack orientation with respect to the rib pattern of the reinforcing bars has little or no effect on the obtained bond properties. Three different crack orientations were tested and the results showed similar ultimate bond strengths.

(3) For double cracked specimens the order in which the cracks are formed (linked to the test method) has no significant influence on the bond behaviour.

(4) Confinement influences the ultimate bond strength of a pre-cracked specimen. In the absence of a restraining force, existing cracks can open up fully enabling the reinforcing bar to more easily slip out of the specimens. The reduction of bond capacity between confined and unconfined specimens was an additional 11% for a given parameter combination.

(5) When the concrete cover is reduced, the residual bond strength after cracking is reduced as well. For smaller covers the failure mode of the uncracked concrete shifts from a pull-out failure to a splitting failure.

(6) The bond strength of reinforcing bars in cracked cylinders embedded in an uncracked concrete ring of 23 mm is 18% lower than uncracked specimens with the same total concrete cover. However, the obtained values are higher than those obtained for concrete with a single crack extending through the entire concrete cover but confined by a plastic tube of 2.1 mm thick.

The obtained test results indicate that the presence of longitudinal cracks can significantly influence the bond behaviour of ribbed reinforcing bars in concrete. This suggests that bond reduction factors are necessary for cracks that run along the reinforcement bars when undertaking load bearing capacity checks of existing reinforced concrete structures.

ACKNOWLEDGEMENTS

The authors would like to gratefully acknowledge the financial support of the UK Engineering and Physical Sciences Research Council (EPSRC) through the EPSRC Project 'Reinforced concrete half-joint structures: Structural integrity implications of reinforcement detailing and deterioration' [Grant no. EP/K016148/1].

Additional data related to this publication is available at the University of Cambridge's institutional data repository: https://www.repository.cam. ac.uk/handle/1810/248740.

REFERENCES

1. Task Group Bond Models, fib Bulletin 10 – Bond of Reinforcement in Concrete, Lausanne, 2000.
2. BRE Centre for Concrete ConstructionCorrosion of steel in concrete – durability of reinforced concrete structures, Digest 444 (2000) 12.
3. J. Rodriguez, L. Ortega, A. Garcia, Assessment of structural elements with corroded reinforcement, in: Proceedings of the International Conference on Corrosion and Corrosion Protection of Steel in Concrete, 1994, pp. 171–185.
4. C. Andrade, C. Alonso, F. Molina, Cover cracking as a function of bar corrosion: part I-experimental test, Mater. Struct. 26 (1993) 453–464.
5. G. Al-Sulaimani, M. Kaleemullah, I. Basunbul, Influence of corrosion and cracking on bond behavior and strength of reinforced concrete members, ACI Struct. J. 87 (2) (1990) 220–231.
6. L. Clark, M. Saifullah, Effect of corrosion rate on the bond strength of corroded reinforcement, in: Proceedings of the International Conference on Corrosion and Corrosion Protection of Steel in Soncrete, 1994, pp. 591–602.

7. A. Almusallam, A. Al-Gahtani, R. Abdur, Rasheeduzzafar, Effect of reinforcement corrosion on bond strength, Constr. Build. Mater. 10 (2) (1996) 123–129.

8. L. Clark, M. Saifullah, Effect of corrosion on reinforcement bond strength, in: Proceedings of the Conference on Structural Faults & Repair, 1993, pp. 113– 119.

9. C. Alonso, C. Andrade, J. Rodriguez, J. Diez, Factors controlling cracking of concrete affected by reinforcement corrosion, Mater. Struct. 31 (1998) 435– 441.

10. T. Vidal, A. Castel, R. François, Analyzing crack width to predict corrosion in reinforced concrete, Cem. Concr. Res. 34 (1) (2004) 165–174.

11. T. Söylev, R. François, Quality of steel–concrete interface and corrosion of reinforcing steel, Cem. Concr. Res. 33 (9) (2003) 1407–1415.

12. H. Yalciner, O. Eren, S. Sensoy, An experimental study on the bond strength between reinforcement bars and concrete as a function of concrete cover, strength and corrosion level, Cem. Concr. Res. 42 (5) (2012) 643–655.

13. H. Lee, T. Noguchi, F. Tomosawa, Evaluation of the bond properties between concrete and reinforcement as a function of the degree of reinforcement corrosion, Cem. Concr. Res. 32 (8) (2002) 1313–1318.

14. J. Cabrera, P. Ghoddousi, Effect of reinforcement corrosion on the strength of steel concrete bond, in: Proceedings of the International Conference on Bond in Concrete, 1992, pp. 10.11–10.24.

15. D. Coronelli, Bond of corroded bars in confined concrete: test results and mechanical modelling, Stud. E Richerche 18 (1997) 137–211.

16. K. Stanish, Corrossion Effects on Bond Strength in Reinforcement in Concrete, University of Toronto, 1997.

17. G. Mancini, F. Tondolo, Effect of bond degradation due to corrosion – a literature survey, Struct. Concr. 15 (3) (2014) 408–418.

18. P. Gambarova, G. Rosati, B. Zasso, Steel-to-concrete bond after concrete splitting: test results, Mater. Struct. 22 (1) (1989) 35–47.

19. L. Marsavina, K. Audenaert, G. De Schutter, N. Faur, D. Marsavina, Experimental and numerical determination of the chloride penetration in cracked concrete, Constr. Build. Mater. 23 (2009) 264–274.

20. B. Šavija, E. Schlangen, J. Pacheco, R. Polder, Modified Wedge Splitting Test (MWST) – a simple tool for durability investigations of reinforcement corrosion in cracked concrete, in: Concrete Repair, Rehabilitation and Repair III, 2012, pp. 386–391.

21. M. Ismail, A. Toumi, R. François, R. Gagné, Effect of crack opening on the local diffusion of chloride in cracked mortar samples, Cem. Concr. Res. 34 (2004) 711–716.

22. A. Djerbi, S. Bonnet, A. Khelidj, V. Baroghel-Bouny, Influence of traversing crack on chloride diffusion into concrete, Cem. Concr. Res. 38 (2008) 877–883.

23. European Committee for Standardization, EN 1992-1-1 Eurocode 2: Design of concrete structures – Part 1–1: General rules and rules for buildings. European Committee for Standardization, 2004.

24. T.C. Rilem, Bond test for reinforcing steel: 2. Pull-out test, Mater. Struct. 3 (15) (1970) 175–178.

25. P. Desnerck, G. De Schutter, L. Taerwe, Bond behaviour of reinforcing bars in self-compacting concrete: experimental determination by using beam tests, Mater. Struct. 43 (S1) (2010) 53–62.

26. F. de Almeida Filho, M. El Debs, A. de C El Debs, Bond-slip behavior of selfcompacting concrete and vibrated concrete using pull-out and beam tests, Mater. Struct. 41 (6) (2007) 1073–1089.

27. A. Windisch, A modified pull-out test and new evaluation methods for a more real local bond-slip relationship, Mater. Struct. 18 (105) (1985) 181–184.

28. R. Tepfers, A Theory of Bond Applied to Overlapped Tensile Reinforcement Splices for Deformed Bars, 2nd ed., Chalmers University of technology, Goteborg, 1976.

29. European Committee for Standardization, EN 197-1:2011 – Cement – Part 1: Composition, specifications and conformity criteria for common cements. Brussels, Belgium: European Committee for Standardization, 2011.

30. C. Zhang, I. Moore, Nonlinear mechanical response of high density polyethylene. Part I: experimental investigation and model evaluation, Polym. Eng. Sci. 37 (2) (1997) 404–413.

31. European Committee for Standardization, EN 12390–6:2009 – Testing hardened concrete. Tensile splitting strength of test specimens. Brussels, Belgium: European Committee for Standardization, 2009.

2

CHAPTER

A Method for the Assessment of the Internal Structure of Bio-Aggregate Concretes

Joseph Williams, Mike Lawrence, Pete Walker

BRE Centre for Innovative Construction Materials, Department of Architecture and Civil Engineering, University of Bath, Bath BA2 7AY, UK

HIGHLIGHTS

- A novel method for assessing bio-aggregate composites is proposed.
- 2D and 3D image analysis used to assess aggregate orientation.
- Specimens of hemp-lime found to have a strong anisotropic internal structure.

ABSTRACT

The thermal and mechanical properties of bio-aggregate concretes are known to be anisotropic. This is assumed to be the result of an orientated internal arrangement of particles; the internal structure of other aggregate composite materials is known to be a determiner of physical properties and has been the focus of much study. Despite this the internal structure of bio-aggregate concretes has to date only been considered qualitatively. This work presents a novel method for the assessment of the internal structure of bio-aggregate concretes through the application of image analysis. Results are presented for the assessment of hemp-lime specimens and demonstrate a significant anisotropy within the material. These results account for anisotropic thermal and mechanical behaviour observed elsewhere and demonstrate the importance of the internal structure in determining the properties of these materials. This innovative technique represents a significant breakthrough in the search for optimisation of the performance of renewable, low carbon insulation materials. This class of materials is critical to the sustainable future of the construction industry.

Keywords: Bio-aggregate; Hemp-lime; Image analysis; Internal structure

1. INTRODUCTION

Bio-aggregate concretes (BAC) are formed by combining crop based aggregate, mineral binder and water. Hemp-lime, produced from the chopped up stalk (shiv) of the hemp plant, is one of the most best known but other examples include wood chips, sunflower plant stems [1] and flax plant stems [2]. As a result of the photosynthetically used carbon dioxide embodied within the plant materials, BACs have very low embodied carbon [3],[4] and [5] as well as favourable thermal properties for use in building envelopes, including low thermal conductivity and high heat capacity [6], [7], [8] and [9].

To produce BACs the mixed constituents are cast into formwork or spray applied onto a substrate. Both processes apply a unidirectional compacting force to the wet material: tamping or deliberate compaction in the case of cast and force of projection in the case of sprayed. As bio-aggregate particles are often elongated in form, this compacting force is considered by most to have significant influence on their arrangement, the elongated particles tending towards stratified planes that are perpendicular to compacting force [10] and [11].

The anisotropic arrangement concept is supported by observations of some physical properties. Nguyen [12] measured the thermal conductivity of cast-compressed hemp-lime specimens both parallel and perpendicular to the compacting force and found up to a 30% higher thermal conductivity in the perpendicular direction. Other work has reported similar findings for cast-tamped material [13] and sprayed material [14]. In all cases the observations were attributed to the internal structure, in which the stratified planes are opposing the transverse heat transfer.

The compressive behaviour of BACs has also been studied with respect to orientation of the bio-aggregates. Previous studies have observed that the peak compressive strength of these materials is higher when loaded parallel to the compacting force however the compressive stiffness is lower [15], [16] and [17]. Additionally the material has been observed to fail in a more brittle manner when loaded perpendicularly. Figure 1 shows the average compressive stress/strain plots of three 150 mm cube specimens of hemp lime and highlights this difference in behaviour that can be attributed to the internal structure – the strata providing differing transverse and parallel load paths.

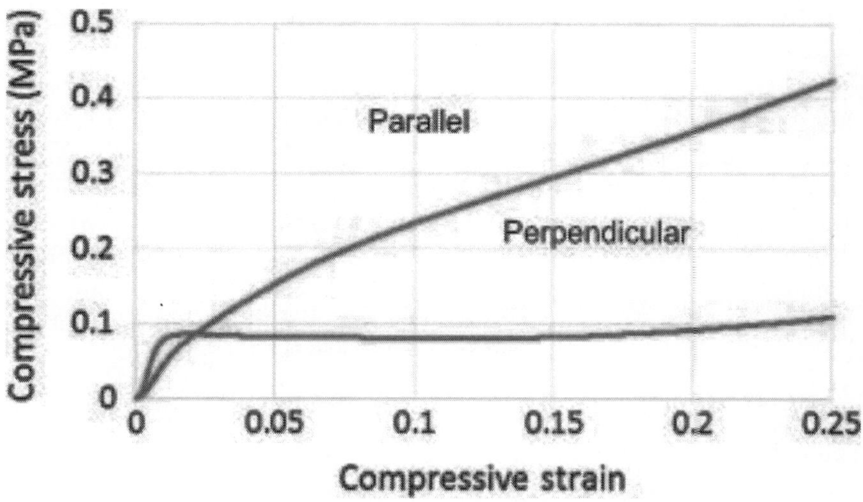

Figure 1. Stress strain plots for perpendicular and parallel loaded hemp-lime cubes produced with a ratio of 16% hemp, 36% binder and 48% water and cast in three layers of 0.66 kg with vertical compaction applied between each layer.

Studies of other particulate composite materials such as asphalt [18] and [19] and soil[20], have demonstrated that the arrangement of the internal structure in these cases is one of the contributing factors for the global properties and a cause of anisotropic behaviour. Despite this, and the observation of anisotropic properties, there has, to the authors' knowledge, been no attempt to measure the internal structure of bio-aggregate concretes, for orientation or for any other parameter.

In order to test the hypothesis of compaction generating strata within bio-aggregate concretes, this paper outlines an image analysis method for the statistical assessment of particle orientation within cast material. The method is based on established methods for other materials as well as image analysis methods already in use for the study of the size distribution of bio-aggregate particles. Results are reported from the application of this method to three variants of the bio-aggregate concrete hemp-lime.

2. METHODOLOGY

2.1. Materials and specimen production

All the specimens considered in this work were produced using hemp shiv grown in the UK the properties of which are presented in Table 1. The particle size distribution of bio-aggregates length and width can be determined by two dimensional digital image analysis as established by others [21]. A 20 g sample of shiv was assessed in this way with the results presented in Figure 2. The results presented in Table 1 and Figure 2 indicate that the material used in this study is comparable to that used widely elsewhere in the literature and that the shiv particles can be considered to be of elongated form.

Table 1. Properties of hemp shiv.

Bulk density (kg m^{-3})	Particle length (mm)		Particle width (mm)	
	Mean particle length (mm)	Standard deviation	Mean	Standard deviation
108	5.56	5.49	1.55	1.33

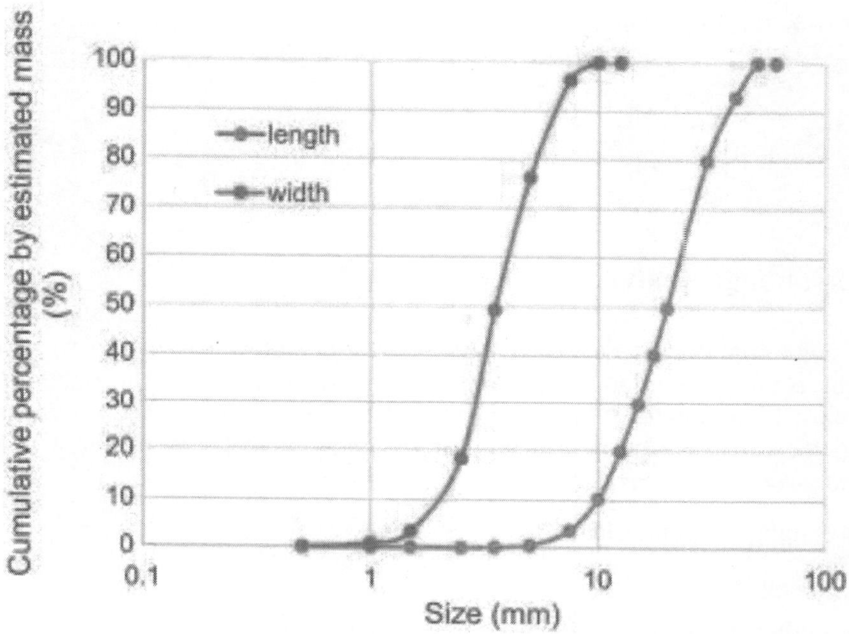

Figure 2. Particle size distribution of hemp shiv as obtained by two dimensional image analysis of a 20 g sample.

The binder used was pre-formulated for hemp-lime construction. The specific formulation of the binder is not declared by the manufacturer, however it is believed to be mainly hydrated lime with additional pozzolanic additives and cement.

To produce the composite material the constituents were mixed in a revolving pan mixer by first slaking the binder with the water and then adding the hemp and further mixing until uniformly combined. For each specimen a set mass of material was tamped into a 150 mm cube mould fitted with a collar to allow for initial overfilling. Three variations of material, standard, light weight and compact, were used in the study Table 2. The standard mix represents an industry standard hemp-lime formulation for walling applications. The light weight mix represent a material with the same proportion of hemp per unit volume but a lower binder content; the compacted mix represents a higher density version of the standard mix but with the same binder to aggregate ratio.

Table 2. Hemp-lime mixture variations.

Mixture	Aggregate (bulk%)		Binder (bulk%)		Water (bulk%)		Paste (%)		Wet mass of specimen (kg)	Estimated Porosity (Fresh state%)	Average dry density (kgm^{-3})
	Mass	Vol	Mass	Vol	Mass	Vol	Mass	Vol			
Standard	16	56	36	23	48	21	84	32	2.00	66	284
Light weight	21	64	36	20	43	16	79	25	1.48	84	353
Compacted	16	56	36	23	48	21	84	32	2.41	60	422

2.2. Image analysis

Image analysis is the process of imaging a material either in two or three dimensions and then using analysis conducted on the image to extract meaningful data. In this application the required outcome was to gauge if a preference of orientation is present in the particles of the material and the process utilised was as follows:

- image the material;
- segregate out the shiv;
- identify and measure the orientation of the particles;
- perform a frequency analysis of the population.

Similar methods, but for the study of other materials, have already been developed and were used as a reference for application to bio-aggregate concretes. These include asphalt [22] and [23], soil [20], fibre reinforced concrete [24] and [25] and fibre board[26] and [27]. Image analysis has also been applied to investigate the size distribution of bio-aggregate particles [21] and was used as an additional reference. Two methods of imaging were considered within the study in order to provide comparison and validate results: visual digital imaging (VD imaging) and computer tomography scanning (CT scanning).

VD imaging is conducted by taking physical sections of the material and imaging the exposed face using a digital camera or flat-bed scanner. An advantage of this method is its low cost however it is a destructive method and requires imaging in several orientations in order to assess the three dimensional topology. Further disadvantages specific to use with BACs are the similar colouring of the particles and binder and the large number of voids that makes visual differentiation of components difficult.

In this study, pairs of specimens were VD imaged, one parallel to compacting force and one perpendicular. Each specimen was sectioned into six 150 mm square slices using a band saw. To improve the visual contrast between material and voids, the slices were submerged in a low viscosity coloured resin. Once cured, the faces were sanded to remove excess and scanned with a flatbed scanner at a resolution of 2400 dpi. To improve contrast between the binder and the hemp, a coloured pigment was added to the binder during the specimen production at a dose of 10 g per specimen.

CT scanning is an established method of imaging the internal structure of materials producing a map of the X-ray absorbance within the material, a property closely linked to density. Advantages of this method are that it is non-destructive and produces a three dimensional image of the material. It is however expensive in comparison to VD imaging and the contrast within the parameters it measures cannot be enhanced by physical pre-treatments.

Within this study all CT scans were taken using a Nikon XTEK, XTH 225 ST CT scanner set to 165 kV and 165 μA. The data were recompiled into a three dimensional volume using Avizo® Fire 8.

2.3. Segregation

To identify the shiv within the images, two image filters with successful precedents for other materials [23], [28] and [29] were applied: a median filter and a threshold filter. The same processes were implemented for both two dimensional and three dimensional images, however differing programs and settings were used due to the differing image formats. ImageJ and Avizo® Fire 8 were used for the two dimensional and three dimensional images respectively.

A median filter replaces each two dimensional pixel's, or three dimensional voxel's, value with the median of those within a specified radius [30] and [31]. This has the result of smoothing the image and removing anomalies Figure 3. A threshold filter is used to segregate the smoothed image. Each pixel/voxel is in turn assessed against a set of criteria in order to produce a binary image [30] and [31]. Careful selection of the criteria enables the identification of a single component and in this case was used to identify the shiv Figure 3.

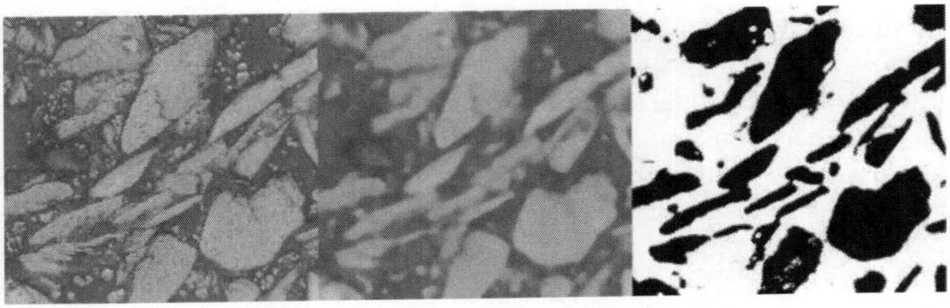

Figure 3. Left to right: original VD image, image passed through median filter, image passed through threshold filter.

Both the median filter and threshold filter applied by each program are controlled by operator assigned settings. The nature of these settings differs between the two programs used, however, in both cases their selection will have a bearing on the analysis results. To assess the sensitivity of the analysis to the choice of settings used, a sensitivity analysis was undertaken on a single image from each imaging method. The range of settings considered for each of the imaging methods is presented in Table 3.

Table 3. Image enhancement and filtering settings trialled, those deemed most appropriate identified in bold.

Process	ImageJ settings	Avizo® Fire 8 settings
Median filter	Radius (pixels): 5, 10, 15, **20**, 25, 30	Iterations: 5, **6**, 7
Colour threshold lower boundary	8 bit hue value: 10, 12, **14**, 16, 18, 20	32 bit float intensity value: **4**, 5, 6 (basic mix) 4, 4.5, **5** (light mix) 6, **7**, 8 (compacted mix)
Colour threshold upper boundary	8 bit hue value: **50**	32 bit float intensity value: 28, **29**, 30 (basic mix) 8, 8.5, **9** (light mix) 28, **29**, 30 (compacted mix)
Opening algorithm	Iterations: 1, 2, **3**, 5, 10, 15, 20, 25	Radius (voxels) 1, 2, **3**

2.4. Identification and measuring orientation

Within the binary images, particles are considered to be regions of interconnected positive pixel/voxels. Due to the close proximity of particles and the limited resolution of the images, there are regions where several aggregates may be misinterpreted as a single particle due to touching areas. To help mitigate the extent of misinterpreted particle groups, an opening algorithm was applied to the images prior to final assessment Figure 4.

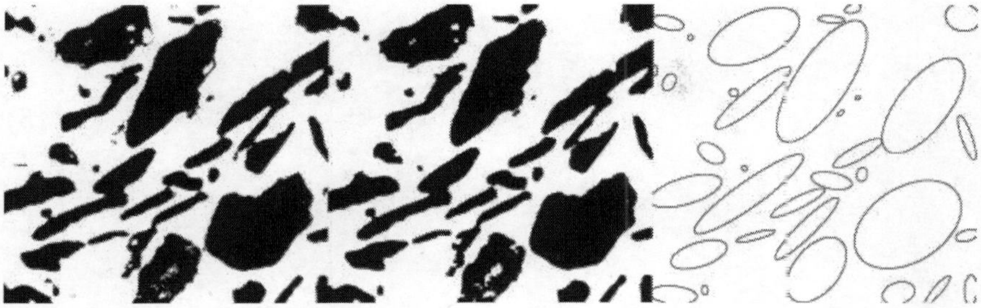

Figure 4. Left to right: thresholded image, image passed through opening algorithm, image assessed for particles and represented as best fitting ellipses.

Opening algorithms selectively remove and then add additional positive pixels or voxels based on the values of those around it [30] and [31]. These are commonly used at this stage of image analysis in order to clean the images and help remove bridges between adjoining particles. As the opening algorithm in both cases was controlled by a user assigned setting, again a range of settings was trialled with details given in Table 3.

To measure the orientation of the identified particles, inbuilt tools within the respective programs were used. The process in both cases is based on calculation and interpretation of the first and second order image moments and is detailed fully elsewhere [32] and [33]. For each group of *n* connected pixels/voxels, the global first and second moments are calculated:

$$m_{pqr} = \sum_{i=1}^{n} x_i^p y_i^q z_i^r \tag{1}$$

The central second moments: u_{200}, u_{020}, u_{002}, u_{011}, u_{101}, and u_{110} and the centroid coordinates can then be determined:

$$u_{pqr} = m_{pqr} - \frac{m_{100}^p m_{010}^q m_{001}^r}{m_{000}} \tag{2}$$

$$\bar{x} = \frac{m_{100}}{m_{000}} \quad (3)$$

$$\bar{y} = \frac{m_{010}}{m_{000}} \quad (4)$$

$$\bar{z} = \frac{m_{001}}{m_{000}} \quad (5)$$

By equalising the second moments and centroid of the particle with that of an ellipse or ellipsoid, the best fitting ellipse or ellipsoid can be determined Figure 4. This enables a consistent measurement of geometrical properties of particles despite their irregular shapes. The orientation of the principal axis of the ellipse or ellipsoid was used as the measure of particle orientation and can be found in each 2D plane:

$$\theta_{xy} = \frac{1}{2} \tan^{-1} \left(\frac{2u_{110}}{u_{200} u_{020}} \right) \quad (6)$$

$$\theta_{yz} = \frac{1}{2} \tan^{-1} \left(\frac{2u_{011}}{u_{020} u_{002}} \right) \quad (7)$$

$$\theta_{zx} = \frac{1}{2} \tan^{-1} \left(\frac{2u_{101}}{u_{002} u_{200}} \right) \quad (8)$$

2.5. Frequency analysis

For visual digital imaging, twelve images were produced for each mix: six taken parallel to compaction and six perpendicular. For each mix a cumulative frequency distribution was produced in each direction for the total particle population from all six slices. In the case of CT scanning, one image was taken per mix with two values of particle orientation measured that equate to the same parallel and perpendicular viewpoints. For each mix a cumulative frequency distribution was produced in each direction for the total particle population of the image. In both cases only particles above a minimum size of 1 mm^2 and 1 mm^3 respectively were considered. The data from both imaging methods were manipulated to give an orientation scale of between 0° and 90° which represent horizontal and vertical alignment in the image plane respectively.

3. RESULTS

3.1. Sensitivity to image enhancement parameters

The perpendicular frequency distributions obtained from one image but using the range of parameters provided in Table 3 were compared. In each instance a version of the Hankinson equation [34], Eq. (9), was fitted to the distribution using a least squares approach and the determined values of f_0, f_{90} and n used as indicators by which to compare the distributions. Hankinson's equation was originally developed for modelling the compressive stress of spruce with respects to orientation of the grain but has since been developed for modelling other properties of timber as well as wood based particle composite boards [35], [36] and [37]. Hankinson's equation was selected due to strong similarities between distribution forms observed in this study and those where the equation has been previously applied. In addition the boundary criteria, f_0 and f_{90}, enable the distribution to be applied both to the 2D images, where a lower bound frequency can be observed, as well as the 3D images. To the author's knowledge Hankinson's equation has not previously been used to model frequency of orientated particles/fibres however the wrapped Cauchy distribution that can be manipulated into the same "Hankinson" form has been used successfully [38].

$$f = \frac{f_0 f_{90}}{f_0 \sin^n \theta + f_{90} \cos^n \theta} \tag{9}$$

The values of n, for differing enhancement parameters, are presented for the standard mix in Figs. 5 and 6. for two dimensional and three dimensional images respectively. The values of f_0 and f_{90} with respect to the enhancement parameters, while not shown, exhibit similar patterns and the same pattern was found in all three mix designs.

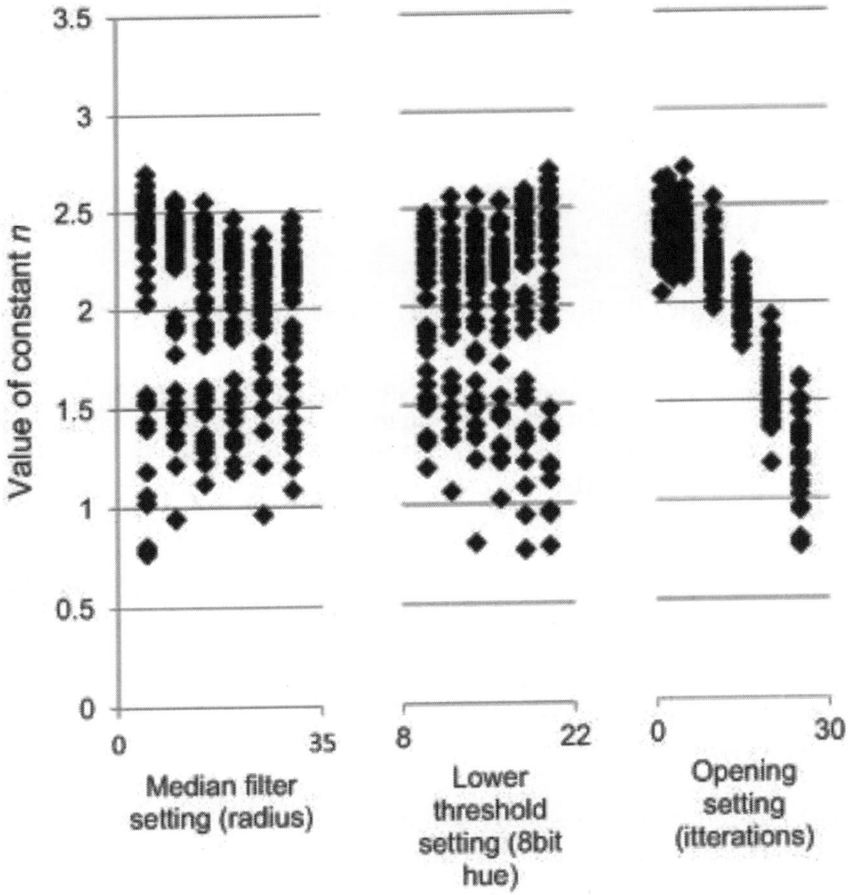

Figure 5. The value of the constant n obtained using differing enhancement setting with 2D images.

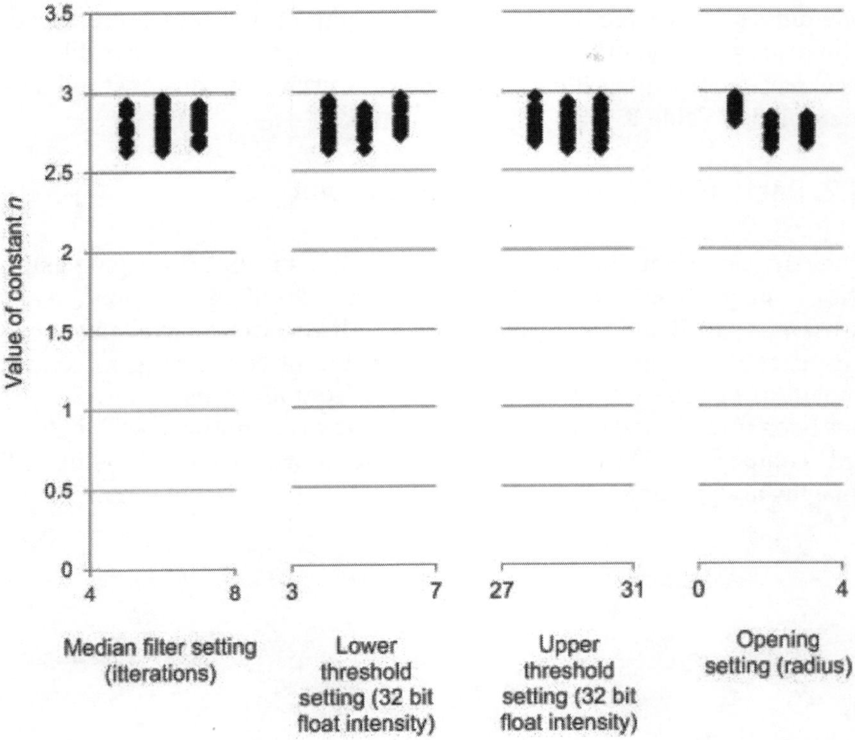

Figure 6. The value of the constant n obtained using differing enhancement setting with 3D images.

In Figs. 5 and 6 it is evident that the enhancement parameters can have a significant influence on the value of n and therefore the distribution produced; for the visual digital imaging method it was determined that the range of parameters considered was able to effect a 10% variation in the peak frequency of the distribution. The opening algorithm can be seen to be the most onerous process and the smaller number of opening algorithm variations considered in the CT data is the likely reason for the smaller spread of results in this case. When the opening algorithm iterations are restricted to under three iterations for the digital visual image data, the variation in peak frequencies is reduced to 2.4% and is comparable to the variation in the CT images.

In order to select appropriate parameters to use for the full assessment, the enhanced images from the sensitivity study were compared against the original images by eye. This was conducted using a "winner stays on" system whereby two images were compared to the original with the one judged least effective at identifying hemp particles being replaced. The process was continued until all permutations had been considered. The parameters identified as most appropriate by this method are identified in Table 3.

3.2. Particle orientation frequency analysis

The particle orientation frequency distributions from VD imaging using the enhancement and filtering setting identified in Table 3 are presented in Figure 7. The resulting distributions, viewing both perpendicularly and parallel to the direction of compacting force, are shown as are the distributions produced from all three mixes. Eq. (9) has been fitted to the data in using a least squares approach each case to aid comparison. The equivalent distributions produced using CT imaging are presented in Figure 8.

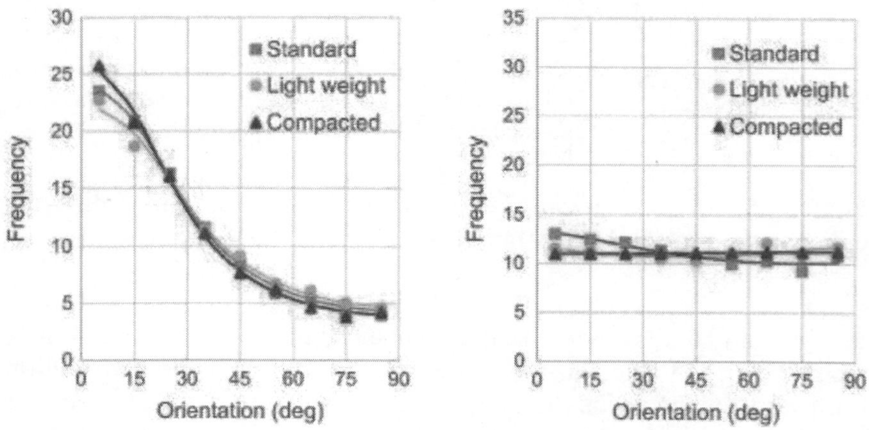

Figure 7. Frequency distribution of particle orientation produced using two dimensional imaging. Perspective perpendicular to compaction, left, and parallel to compaction, right.

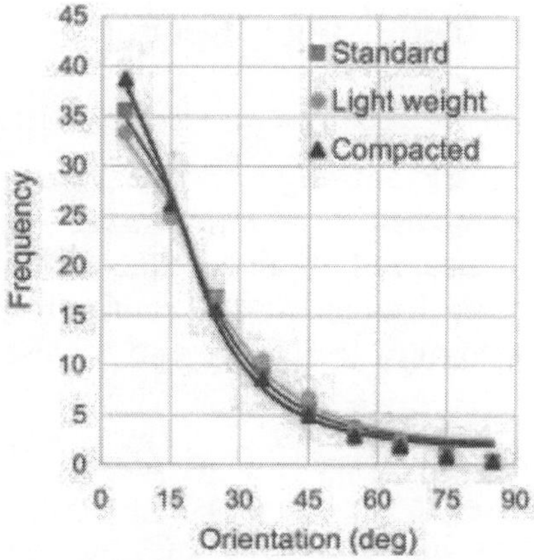

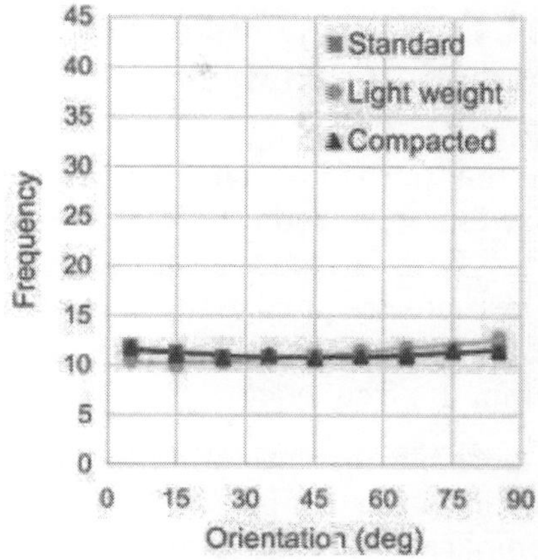

Figure 8. Frequency distribution of particle orientation produced using CT imaging. Perspective perpendicular to compaction, left, and parallel to compaction, right.

4. DISCUSSION

4.1. Comparison of material in opposing directions

From Figs. 7 and 8 it can be seen that the frequency distribution of particle orientations viewed parallel to compacting force is flat: indicative of a random distribution and suggests the structure is isotropic in this plane. Conversely, in the perpendicular plane, the frequency distribution of particle orientations is strongly swayed. This indicates a preference of orientation within the population and is therefore suggestive that the material's structure is anisotropic in this plane.

The finding of both an anisotropic and isotropic distributions in contrasting directions is supportive of the hypothesis that compaction force produces a preferential orientation [10] and [12]. A clear tendency is observed in all specimens for the major axis of the particles to tend towards horizontal planes, perpendicular to the compaction force. This observation concurs with, and provides an explanation for the observations of others that several physical properties of bio-aggregate composites exhibit anisotropic behaviour with respect to the direction of compaction force [15], [16] and [17].

4.2. Comparison of materials produced using differing formation variables

In the perpendicular perspective, a consistent difference between distributions produced from the differing mixes is present in the results from both methods. The light mix is observed to have the lowest maximum frequency and highest minimum frequency; the compacted mix is observed to have the highest maximum frequency and lowest minimum frequency.

A possible interpretation is that the overall degree of orientation is higher in the compacted material and lower in the low density material compared to the standard mix. This is logical in the case of the compacted mix where the ratio of constituents was the same but level of compaction was known to be higher. For the light weight mix, the density of hemp in the material was controlled to be the same as the standard mix and this was assumed to indicate a similar compaction level and thus the cause of the apparently lower degree of orientation in the case of the lightweight mix can therefore not be attributed to compaction. Alternately it is proposed that the aggregate volume fraction may be a determining factor in the degree of orientation; a lower volume of aggregates producing a higher level of inter-particle porosity and increasing the ability of particles to rotate under compaction. It is considered likely therefore that both compaction and aggregate volume fraction will impact the degree of orientation and further experiments to investigate the impact of both variables on the internal structure are proposed.

The observed difference between the frequency distributions from the three mixes is noted to fall within the difference obtained using differing image enhancement settings and highlights the importance of using consistent settings for comparing material. A consistent set of enhancement settings was used across all specimens analysed with VD imaging and so removes this as a source of uncertainty in this case. The variation observed in the perpendicular perspective was also noted to be of greater magnitude than those observed in the parallel perspective. This indicates that the observations in the perpendicular perspective are within the natural variation of the population and therefore the observation that there are differing degrees of orientation, while promising, is statistically insignificant and undermines any firm conclusions at this stage.

4.3. Comparison of digitisation methods

The frequency distributions produced by the VD imaging Figure 7, are observed to be of the same general form of those produced using CT images Figure 8. In each case the parallel perspective has produced a relatively level distribution with all frequencies approximately between 9% and 14%. In the perpendicular perspective both methods have indicated a swayed distribution towards the horizontal with a similar form in both cases. In both cases the ordering of the mixes, least to most orientated, is the same. A smoother frequency distribution was found using the CT images as opposed to the VD imaging and is attributed to the larger overall population of particles measured.

In the perpendicular view, the CT generated images have, in all variations of material, yielded distributions with a consistently higher frequency range. This difference is attributed to the differing nature of the methods; VD imaging analysis measures a random section through each particle while CT scan analysis measures the whole particle and indicates that a consistent method must be adopted for the comparison of material.

The strong correlation between the results from both methods provides a high level of confidence in the results. VD imaging conducted in two planes is considered sufficient to infer the three dimensional structure of the material and, due to the lower costs, is considered to be the more practical of the two methods for larger samples of material.

5. CONCLUSION

In this work the novel application of image analysis methodology for the assessment of the internal structure of bio-aggregate concretes was trialled. It was shown that similar results are obtainable using two differing methods of imaging: CT scanning and VD imaging. This indicates that both procedures are capable of producing reliably analysable images and that image analysis is a suitable method for the assessment of these materials.

The results in this study support the established theory that these materials have an orientated internal structure that is determined by the direction of any compacting force. A consistent difference was also observed between materials constructed in differing ways and, while this was within the natural variation of the results, indicates that the production method is likely to determine the degree of orientation.

This study demonstrates that image analysis is a reliable and repeatable method of assessing the internal structure of bio-aggregate concretes. This will facilitate the development of models which will better predict the impact of construction processes on the mechanical and hygrothermal properties of these materials, leading to the potential optimisation of their performance. Additional work is required to establish the exact way in which construction processes influence the internal structure of bio-aggregate concretes in order to utilise this aspect of the material.

ACKNOWLEDGEMENTS

The authors gratefully acknowledge the financial support provided by the Engineering and Physical Sciences Research Council – United Kingdom, EP/L016869/1.

REFERENCES

1. V. Nozahic, S. Amziane, G. Torrent, K. Saïdi, H. De Baynast, Design of green concrete made of plant-derived aggregates and a pumice–lime binder, Cement Concr. Compos. 34 (2) (2012) 231–241.

2. A. Korjenic, V. Petránek, J. Zach, J. Hroudová, Development and performance evaluation of natural thermal-insulation materials composed of renewable resources, Energy Build. 43 (9) (2011) 2518–2523.

3. M. Boutin, C. Flamin, S. Quinton, G. Gosse, Analyse du Cycle de Vie de Mur en Béton Chanvre Banché sur Ossature en Bois, INRA, Lille, France, 2005. [4] K. Ip, A. Miller, Life cycle greenhouse gas emissions of hemp–lime wall constructions in the UK, Resour. Conserv. Recycl. 69 (2012) 1–9.

4. S. Pretot, F. Collet, C. Garnier, Life cycle assessment of a hemp concrete wall: Impact of thickness and coating, Build. Environ. 72 (2014) 223–231.

5. L. Arnaud, D. Samri, É. Gourlay, Hygrothermal Behavior of Hempcrete. Bioaggregate-based Building Materials, John Wiley & Sons, 2013, pp. 179–242.

6. F. Collet, S. Pretot, Experimental highlight of hygrothermal phenomena in hemp concrete wall, Build. Environ. 82 (2014) 459–466.

7. M. Lawrence, E. Fodde, K. Paine, P. Walker, Hygrothermal performance of an experimental hemp-lime building, Key Eng. Mater. 517 (2012) 413–421.

8. F. Collet, S. Prétot, Thermal conductivity of hemp concretes: variation with formulation, density and water content, Constr. Build. Mater. 65 (2014) 612– 619.

9. Y. Hustache, L. Arnaud, Synthèse des connaissances sur les bétons et mortiers de chanvre, Construire en Chanvre, 2008.

10. S. Elfordy, F. Lucas, F. Tancret, Y. Scudeller, L. Goudet, Mechanical and thermal properties of lime and hemp concrete ("hempcrete") manufactured by a projection process, Constr. Build. Mater. 22 (10) (2008) 2116–2123.

11. T.T. Nguyen, V. Picandet, P. Carre, T. Lecompte, S. Amziane, C. Baley, Effect of compaction on mechanical and thermal properties of hemp concrete, Eur. J. Environ. Civ. Eng. 14 (5) (2010) 545–560.

12. T. Dinh, C. Magniont, M. Coutand, G. Escadeillas, Hemp concrete using innovative pozzolanic binder, in: First International Conference on Bio-based Building Materials, 2015. Clermont-Ferrand, France.

13. T. Pierre, T. Colinart, P. Glouannec, Measurement of thermal properties of biosourced building materials, Int. J. Thermophys. 35 (9–10) (2014) 1832– 1852.

14. S. Amziane, V. Nozahic, M. Sonebi, Design of mechanically enhanced concrete using hemp shiv, in: First International Conference on Bio-based Building Materials, 2015. Clermont-Ferrand, France.

15. A. Kashtanjeva, M. Sonebi, S. Amziane, Investigation of the mechanical performance and drying shrinkage of hemp concrete, in: First International Conference of Bio-based Building Materials Clermont-ferrand, France, 2015.

16. Youssef, T. Lecompte, V. Picandet, N. Challamel, Compressive and shearing behaviour of lime and hemp concrete, in: First International Conference on Bio-based Building Materials, 2015. Clermont-Ferrand, France.

17. M. Hamzah, W. Von, N. Abdullah, Effects of compactor types on aggregate orientation of asphalt mixtures, Int. J. Eng. Trans. A Basics 26 (7) (2013) 677. E. Sanchez-Alonso, A. Vega-Zamanillo, D. Castro-Fresno, Effect of type of compaction on mechanical properties in warm-mix asphalts, J. Mater. Civ. Eng. 24 (8) (2012) 1043–1049.

18. Shi, Y. Murakami, Z. Wu, Orientation of aggregates of fine-grained soil: quantification and application, Eng. Geol. 50 (1) (1998) 59–70.

19. V. Picandet, Characterization of Plant-Based Aggregates. Bio-aggregate-based Building Materials, John Wiley & Sons, 2013, pp. 27–74.

20. M.E. Kutay, E. Arambula, N. Gibson, J. Youtcheff, Three-dimensional image processing methods to identify and characterise aggregates in compacted asphalt mixtures, Int. J. Pavement Eng. 11 (6) (2010) 511–528.

21. N. Roohi Sefidmazgi, H.U. Bahia, Effect of compaction conditions on aggregate packing using 2-dimensional image analysis and the relation to performance of HMA, Mater. Struct. 47 (8) (2014) 1313–1324.

22. Y. Akkaya, S. Shah, B. Ankenman, Effect of fiber dispersion on multiple cracking of cement composites, J. Eng. Mech. 127 (4) (2001) 311–316.

23. J. Liu, C. Li, J. Liu, G. Cui, Z. Yang, Study on 3D spatial distribution of steel fibers in fiber reinforced cementitious composites through micro-CT technique, Constr. Build. Mater. 48 (2013) 656–661.

24. E. Badel, C. Delisee, J. Lux, 3D structural characterisation, deformation measurements and assessment of low-density wood fibreboard under compression: the use of X-ray microtomography, Compos. Sci. Technol. 68 (7–8) (2008) 1654–1663.

25. G. Standfest, S. Kranzer, A. Petutschnigg, M. Dunky, Determination of the microstructure of an adhesive-bonded

medium density fiberboard (MDF) using 3-D sub-micrometer computer tomography, J. Adhes. Sci. Technol. 24 (8–10) (2010) 1501–1514.

26. Bessa, V. Castelo Branco, J. Soares, Evaluation of different digital image processing software for aggregates and hot mix asphalt characterizations, Constr. Build. Mater. 37 (2012) 370–378.

27. A.R. Coenen, M.E. Kutay, N.R. Sefidmazgi, H.U. Bahia, Aggregate structure characterisation of asphalt mixtures using two-dimensional image analysis, Road Mater. Pav. Design 13 (3) (2012) 433–454.

28. FEI Visualization Sciences Group, Avizo 8, Avizo User's Guide, English, KonradZuse-Zentrum für Informationstechnik, Berlin, Germanry, 2014.

29. T. Ferreira, W. Rasb, ImageJ user guide, 2012.

30. H. Ming-Kuei, Visual pattern recognition by moment invariants, IRE Trans. Inform. Theory 8 (2) (1962) 179–187.

31. M.R. Teague, Image analysis via the general theory of moments/ , J. Opt. Soc. Am. 70 (8) (1980) 920–930.

32. R. Hankinson, Investigation of crushing strength of spruce at varying angles of grain, Air Serv. Inform. Circ. 3 (259) (1921) 130.

33. H. Holmberg, Influence of grain angle on Brinell hardness of Scots pine (Pinus sylvestris L.), Holz als Roh- und Werkstoff 58 (1) (2000) 91–95.

34. S. Kazemi Najafi, A. Abbasi Marasht, G. Ebrahimi, Prediction of ultrasonic wave velocity in particleboard and fiberboard, J. Mater. Sci. 42 (3) (2007) 789–793.

35. T. Nishimura, P.M. Ansell, Monitoring fiber orientation in OSB during production using filtered image analysis, Wood Sci. Technol. 36 (3) (2002) 229–239.

36. K. Schulgasser, Fibre orientation in machine-made paper, J. Mater. Sci. 20 (3) (1985) 859–866.

3

CHAPTER

Investigation on the Effectiveness of Aqueous Carbonated Lime in Producing an Alternative Cementitious Material

Byung-Wan Jo, Sumit Chakraborty, Ji Sun Choi, Jun Ho Jo

Department of Civil and Environmental EngineeringHanyang University

ABSTRACT

With the aim to reduce the atmospheric CO_2, utilization of the carbonated lime produced from the aqueous carbonation reaction for the synthesis of a cementitious material would be a promising approach. The present investigation deals with the aqueous carbonation of slaked lime, followed by hydrothermal synthesis of a cementitious material utilizing the carbonated lime, silica fume, and hydrated alumina. In this study, the aqueous carbonation reaction was performed under four different conditions. The TGA, FESEM, and XRD analysis of the carbonated product obtained from the four different reaction conditions was performed to evaluate the efficacy of the reaction conditions used for the production of the carbonated lime. Additionally, the performance of the cementitious material was verified analyzing the physical characteristics, mechanical property and setting time. Based on the results, it is demonstrated that the material produced by the hydrothermal method possesses the cementing ability. Additionally, it is revealed that the mortar prepared using the alternative cementitious material yields 33.8 ± 1.3 MPa compressive strength. Finally, a plausible reaction scheme has been proposed to explain the overall performances of the aqueous carbonation as well as the hydrothermal synthesis of the cementitious material.

Keywords: carbon dioxide emission aqueous carbonation CO_2 sequestration hydrothermal synthesis cementitious material

1 INTRODUCTION

Nowadays, the world's increasing need is to control the global warming and environmental pollution minimizing the emission of the climate changing gasses (Jacobsen et al.2013; Phair 2006). It is stated in the Kyoto protocol 6 that including the carbon dioxide, some other gasses such as methane, nitrous oxide, sulfur dioxide, sulfur hexafluoride, hydrocarbons, chlorofluorocarbons, and perfluorocarbons, etc., are responsible for the greenhouse effect and global warming. Usually, the fossil fuel is considered to be a major energy source worldwide, which emits a massive carbon dioxide during burning (Siegenthaler and Oeschger 1987; Keeling et al. 1995; Olajire2013). Conversely, the cement production method emits an enormous amount of carbon dioxide. It is reported elsewhere that the production of the 1-ton cement emits ~830 kg of CO_2 (Amato 2013; Schrabback 2010). Additionally, it is also reported that the level of the CO_2 has increased ~30 % since the beginning of the industrial revolution (Siegenthaler and Oeschger 1987; Keeling et al. 1995). The current level of the atmospheric carbon dioxide (CO_2) is measured to be 400 ppm (Amato 2013; Schrabback 2010), which is projected to be increased up to ~800 ppm by the end of this century without taking a courageous step (Feely et al. 2004). Hence, an immediate practical plan is required to reduce the CO_2 emission. Additionally, the awareness of the people needs to be increased in controlling the mentioned threats. Therefore, it is essential to develop an alternative technology for minimizing the CO_2 released by man-made and industrial activities.

With the aim to reduce the atmospheric CO_2, the attention of the scientists and the technologists was attracted towards the mineral carbonization. The capture or storage of CO_2 in the geological form is one of the most promising approaches to reduce the atmospheric CO_2 (Gerdemann et al. 2007; Chizmeshya et al. 2007; Chen et al. 2006). The numerous investigations were executed to reduce the atmospheric carbon dioxide by in-situ or the ex-situ mineral sequestration method.

Mineral carbonation process can able to minimize the atmospheric CO_2 transforming into the stable carbonate minerals, such as calcite ($CaCO_3$), dolomite ($CaMg(CO_3)_2$), magnesite ($MgCO_3$) and siderite ($FeCO_3$) (Metz et al. 2005). The mineral carbonation is reported to be an exothermic process (Olajire 2013). It takes place at normal temperature. Equation 1 shows the chemistry of the mineral carbonation reaction.

$$MO + CO_2 \rightarrow MCO_3 + (\text{Heat}) \quad (1)$$

Although, an adequate research report is available related to the mineral carbonation (Huijgen and Comans 2005; Huijgen et al. 2007; Hanchen et al. 2008; Huijgen et al.2006, 2004), however, a very few of those techniques are utilized practically. Recently, some other processes such as the precipitated calcium carbonate formation, carbonation using brines, accelerated weathering of limestone and straightforward carbonation are also reported to consume the atmospheric CO_2 (Chen et al. 2006; Huijgen and Comans 2005; Hanchen et al. 2008). Additionally, the carbonation of concrete, as well as the calcium hydroxide, is also reported to reduce the atmospheric CO_2 (Galan et al. 2010; Han et al. 2011). Currently, some investigations have been executed to reduce the CO_2 emission during the cement production by replacing the clinker with supplementary materials such as natural pozzolans, fly ash, and slag, etc (Schrabback 2010; Criado et al. 2010; Juenger et al. 2011; Amato 2013; Dinakar et al. 2013; Jeon et al. 2015; Kotwal et al. 2015; Roychand et al. 2016). Additionally, some alternative cementitious materials are produced by the alkali activation of the pozzolanic minerals such as ground granulated blast furnace slag, metakaolin and fly ash, etc (Dinakar et al. 2013; Kim et al. 2013; Kar et al. 2014; Jeon et al. 2015; Chindaprasirt and Cao2015). Moreover, it is reported that an alternative cementitious material can be produced by the hydrothermal method utilizing geomaterials to reduce the CO_2 emission during the cement production (Jo et al. 2014a, 2015). The synthesized cementitious material reveals benefits in controlling the physical and mechanical performances of the mortar and concrete. In fact, the geopolymers are composed of aluminosilicate minerals, which has the ability to be hard and sturdy, eventually, lead to gain strength (Xu and Van Deventer 2000; Kar et al. 2014). Although an adequate attention has been paid to reduce the CO_2 emission. However, the utilization of the carbonated material in producing the cementitious material has yet to be investigated adequately. Very recently, we have reported the hydrothermal synthesis of a carbon dioxide-stored cementitious material in controlling the performances of the mortar (Jo et al. 2015).

Reviewing the literature, it is apparent that the utilization of the aqueous carbonated slaked lime for the production of a cementitious material has not studied yet. Hence, In this investigation, we have studied the aqueous carbonation of the slaked lime, followed by hydrothermal synthesis of a cementitious material utilizing the carbonated lime infused with hydrated silica and alumina. The overall process is expected to be beneficial in controlling the atmospheric CO_2. Although the aqueous carbonation of slaked lime can not reduce the ultimate CO_2 emission, nevertheless, it would transform the atmospheric CO_2 to a stable material.

2 EXPERIMENTAL SECTION

A systematic experimental program was arranged for the aqueous carbonation of slaked lime and the hydrothermal synthesis of an alternative cementitious material. The details of the experimental procedure and characterization techniques are described clearly in the succeeding sections.

2.1 Aqueous Carbonation of the Slaked Lime

The slaked lime ($Ca(OH)_2$) powder (98 % pure) was used as a primary reactant for the carbonation reaction. The used slaked lime contains some other oxide materials such as $MgO < 0.4$ %, $SiO_2 < 0.8$ %, $Al_2O_3 < 0.5$ % and $Fe_2O_3 < 0.3$ %. An aqueous solution of the slaked lime was prepared using 0.865 g of dry slaked lime powder and 500 ml of distilled water in a beaker. It is reported elsewhere that the solubility of the hydrated lime in water is ~0.173 at 20 °C. Therefore, in this study, 0.173 % slaked lime solution was prepared. In this investigation, the carbonation reaction was performed in a closed reactor. Figure 1 represents a schematic diagram of the reactor used in this investigation. In the reactor, a mass flow meter (MFC-100) was installed to control the carbon dioxide injection speed as well as to measure the mass of the injected CO_2. Additionally, the instrument was associated with a pH measurement device (pHI-201), which was employed to record the change of pH during the carbonation reaction. After preparing the aqueous slaked lime solution, the beaker was placed into a Teflon made closed drum of the reactor to avoid the contamination of the other gasses and unwanted carbonation. Subsequently, the CO_2 gas was purged into the reacting vessel for 20 min maintaining the temperature of the reaction system 21 °C. In this study, the aqueous carbonation of the slaked lime was performed in four different conditions varying the agitation speed (viz., 200 and 450 RPM) and CO_2 injection speed (0.3 and 0.5 l/min). Table 1 presents the details of the reaction conditions used in this study. After completion of the reaction, the precipitates were filtered off using a Whatman 41 filter paper and an aspirator. Thereafter, the moisture was detached from the precipitates using a freeze dryer. Finally, the carbonated product was stored in a vacuum desiccator and allowed to characterize. In this investigation, the carbonation reaction was repeated three times to obtain an average result.

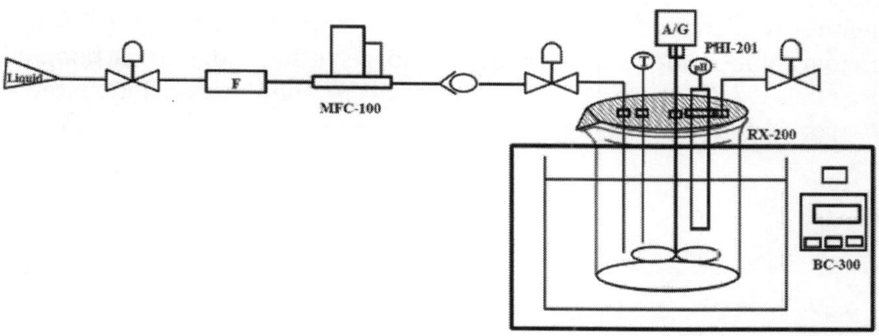

Figure 1. Schematic presentation of the reactor for the aqueous carbonation reaction.

Table 1. Summary of the conditions used for the aqueous carbonization of the slaked lime.

Reaction condition code	Agitation speed (RPM)	Temperature (°C)	CO_2 injection speed (l/min)	Reaction time (min)
A	200	21	0.3	20
B	200	21	0.5	20
C	450	21	0.3	20
D	450	21	0.5	20

2.2 Hydrothermal Synthesis of the Cementitious Material

The carbonated lime produced from the aqueous carbonation method was used for the hydrothermal synthesis of an alternative cementitious material. Initially, 180 g of silica fume was mixed with the 700 ml of water in a beaker using a magnetic stirrer, followed by mixing with 120 g of carbonated lime. Simultaneously, 60 g of sodium hydroxide dissolved in 700 ml of water in an another beaker, followed by mixing with 100 g of sodium aluminate. Afterward, 140 g of triethanolamine (TEA) was added to the sodium aluminate solution for preventing the coagulation of aluminium hydroxide from the solution. Thereafter, the mixtures were agitated for 24 h. A thick gel was produced from the mixture of silica fume and carbonated lime. Conversely, a comparatively thin gel was produced from the hydrated alumina and TEA mixture. Subsequently, the thick gel was added gently to the low-density gel with a constant stirring, followed by 3 h sonication. The mixed compound, thus, obtained was allowed to dry and crystallize in an oven at 105 °C. Finally, the solid crystalline material obtained after 7 days of controlled heating was treated as an alternative cementitious material (ACM). A pictographic model for the hydrothermal synthesis of the carbonated lime based ACM is presented in Figure 2. The solid oven-dried material was, then, ground to make a powder for preparing the mortar samples.

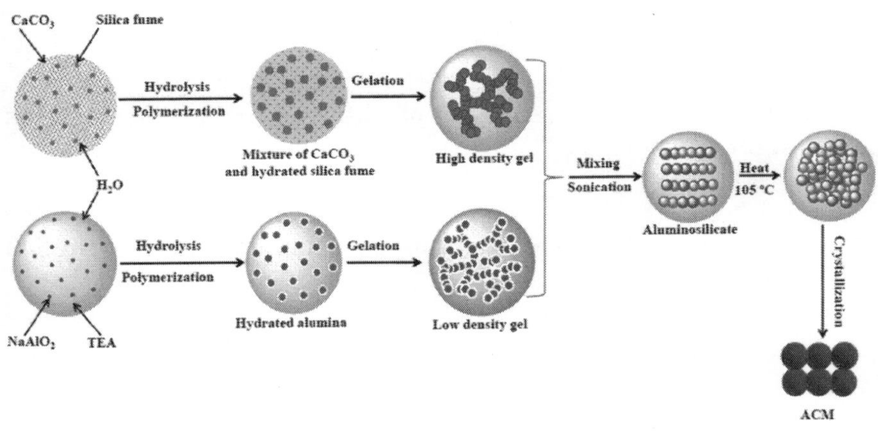

Figure 2. The plausible model for the hydrothermal synthesis of the cementitious material utilizing carbonated material produced by the aqueous carbonation reaction.

2.3 Fabrication of the ACM Based Mortar Sample

In this study, the mortar samples were fabricated using the ACM, fine aggregate (sand), and the varying amounts of alkali activator (50 % NaOH solution) and water. For a particular batch mixing of the ACM based mortar, 100 g of ACM was mixed with the 300 g of fine aggregate, followed by mixing with the required amount of alkali activator and water. In this investigation, a control mortar sample was also prepared using 100 g ordinary Portland cement, 300 g of sand and 60 ml of water. Table 2 shows the mix proportion of the components used for the fabrication of the control and ACM based mortar. The mortar mixture obtained after mixing of the components was, then, cast immediately in the 50 mm side cubic molds and allowed to set for 24 h. After complete setting, the mortar samples were allowed to water cure for 28 days. The mortar samples were removed from the curing chamber and allowed to test after completion of the desired curing period.

Table 2. Formulation code and the mix proportions of the components for the fabrication of the control cement mortar and ACM based mortar.

Sample code	Mix proportioning of the components			
	Cement (g)	Fine aggregate (g)	Water (ml)	Alkali activator (ml)[c]
CCM	100[a]	300	60	–
ACM-M1	100[b]	300	60	–
ACM-M2	100[b]	300	30	30
ACM-M3	100[b]	300	–	60

[a]Ordinary portland cement.

[b]Alternative cementitious material.

[c]50 % NaOH solution.

2.4 Characterizations

The thermal analysis of the precipitates obtained from the aqueous carbonation reaction was performed using a thermogravimetric analyzer (TG-DTA 2020SA). In this experiment, exactly 10 mg of the powder sample was used for the analysis. The experiment was conducted under a dynamic N_2 atmosphere in the temperature range between 30 and 900 °C with a heating rate of 10 °C/min. In this investigation, each sample was analyzed three times to get an average result.

The X-ray diffraction pattern of the precipitates obtained from the aqueous carbonation reaction was recorded using an X-ray diffractometer (Ultima III). The instrument produced a monochromatic X-ray beam of the wavelength 1.54 Å using CuKα radiation (40 kV, 40 mA) and Ni filter. In this study, the X-ray diffraction pattern of the samples was recorded in the 2θ range between 10°–60°, maintaining a scan speed of 1 °/min with a step difference of 0.02″. Prior to analyzing the X-ray diffraction pattern, the sample was packed into a rectangular hollow area of the glass made sample holder, followed by placing in the instrument.

The microstructure of the precipitates obtained from the aqueous carbonation reaction, the product obtained from the hydrothermal synthesis and the 28 days cured mortar was recorded using a field scanning electron microscope (JEOL JSM-6700F). Prior to analyzing the microstructure, a very thin gold was sputter coated on the surface of the moisture-free dried samples to avoid charging. Thereafter, samples were placed on the SEM stub and allowed to analyze. The digital scanning electron micrographs were recorded in 10–20 kV accelerated voltage and 15 k× magnification.

The particle size analysis of the cementitious material obtained from the hydrothermal method was performed using LA-950 Laser particle size analyzer. For the particle size analysis, exactly 1 g of the oven-dried sample was fed into the PowderJet Dry Feeder of the LA-950 Laser particle size analyzer. Furthermore, the sample was analyzed based on the Mie scattering theory. In this instrument, two light sources viz., 5 mW, 650 nm red laser diode and 3 mW, 405 nm blue LED were used to analyze the particle size of the material. In the measurement array, the high-quality photodiodes were used to detect the scattered light over a wide range of angles.

The initial and final setting time of the alternative cementitious material obtained from the hydrothermal method was measured according to the KSL 5108 (2007). Usually, this standard method is used to measure the setting time of the hydraulic cement.

The compressive strength of the 50 mm side cubic alternative cementitious material (ACM) based mortar sample was measured using a universal testing machine (Shimadzu, CCM-200A) with a loading rate 0.06 MPa/min in accordance with the Korean standard KS F 2405 (2010).

3 RESULTS

3.1 pH Analysis of the Carbonation Reaction Medium

Figure 3 depicts the pH variation of the reaction medium with the increase in reaction time. In this study, the pH of the reaction medium was measured at the 1 min interval. From the figure, it is observed that irrespective of the reaction conditions, the pH of the reaction medium decreases gradually up to 5 min of the reaction, followed by keeping constant with the progress in reaction. Prior to the carbonation reaction started (before CO_2 purged), the pH of the reaction medium is measured to be 12.4, which is measured to be ~6.5 after 5 min of the reaction started. Interestingly, the decrement of pH is observed to be different for the different reaction conditions (inset figure of the Figure 3). The decrement of pH for the reaction condition B is observed to be higher as compared to that of the reaction condition D, C and A. Additionally, the order of pH decrement of the four different reaction conditions is found to be B > D > C > A. Hence, It is indicated that the rapid carbonation reaction occurs in the reaction condition B as compared to that of the other three conditions. The pH of the carbonation reaction medium is reduced to be due to the formation of the carbonic acid, which neutralizes the saturated solution of the slaked lime. During the reaction, the lime solution reacts with the carbonic acid, which in turn leads to produce the calcium carbonate and water. As a consequence, the concentration of OH^- in the reaction system decreases. Eventually, the pH of the reaction system decreases (Olajire 2013). The pH of the reaction medium decreases continuously until the lime molecules are available for the carbonation. After complete neutralization of the lime molecules (after 6 min of the reaction started), the solution becomes CO_2 rich (pH = 5–6), which restrains to reduce the pH.

3.2 Quantification of the CO_2 Consumption

Table 3 represents the moles of reactants and products of the aqueous carbonation reaction. In this study, the experiment was repeated three times to identify the most effective reaction condition. After measuring the mass of the reactant and products, the moles of CO_2 consumed in the carbonation reaction in four different reaction conditions are measured. From the table, the moles of CO_2 (average value) consumed in the four

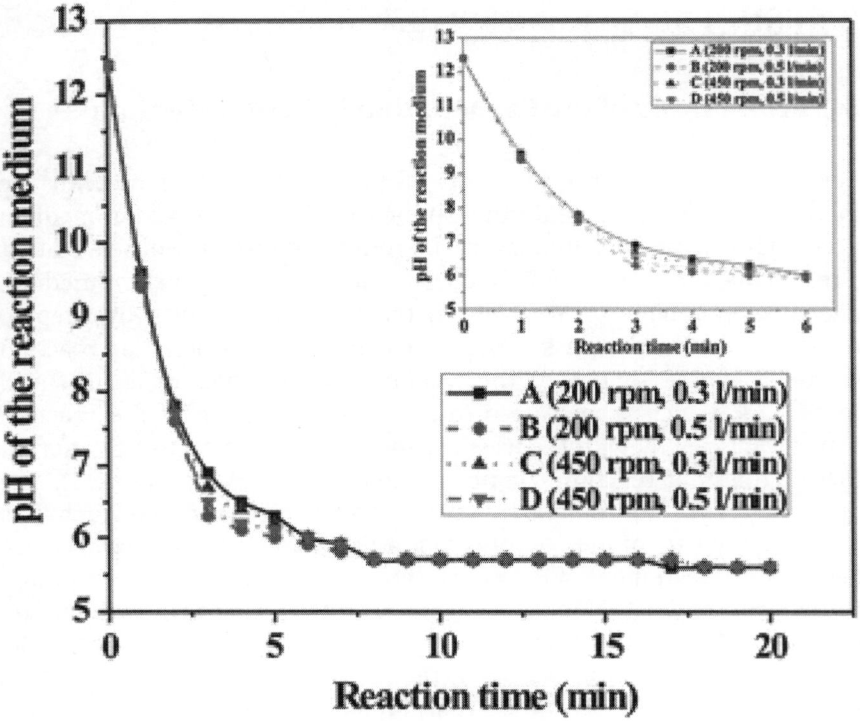

Figure 3. pH variation of the aqueous carbonation reaction medium as function of the reaction time.

different carbonation reaction conditions such as A, B, C and D are measured to be 0.0107 ± 0.0004, 0.0118 ± 0.0008, 0.0100 ± 0.0004, and 0.0112 ± 0.0011, respectively. It is well known that the 1 mol of lime reacts with 1 mol of CO_2 to produce 1 mol of calcium carbonate (Olajire 2013). Viewing in light of the theoretical prediction, it is assumed that the 0.0117 mol of slaked lime would react with 0.0117 mol of CO_2 to produce 0.0117 mol of calcium carbonate. However, in this investigation, all four reaction conditions do not follow the theoretical prediction conveniently. As presented in Table 3, the reaction conditions B and D nearly follow the theoretical prediction. From the table, it is envisaged that in the reaction condition B and D, 0.0117 mol of slaked lime react with 0.0118 ± 0.0008 and 0.0112 ± 0.0011 mol of CO_2, respectively, and produces 0.0116 ± 0.0003 and 0.0111 ± 0.0002 mol of calcium carbonate, respectively.

Additionally, the % of CO_2 consumption in four different reaction conditions such as A, B, C and D is calculated to be 91.2 ± 3.0, 100.1 ± 6.8 85.5 ± 3.1 and 96.0 ± 9.1 %, respectively. Figure 4 represents the moles of slaked lime (calcium hydroxide) and CO_2 used and the moles of calcium carbonate produced in the theoretical carbonation reaction scheme and the four different carbonation reaction conditions such as A, B, C and D. The result reveals that the reaction condition B consumes higher amounts of CO_2 and produces a greater amount of calcium carbonate as compared to that of the reaction condition D, A and C. Moreover, the reaction condition B follows the theoretical prediction more appropriately. Although, the impact of the reaction condition D is similar to the reaction condition B. However, the reaction condition D (450 RPM agitation speed and 0.5 l/min CO_2 injection speed) would be a more energy consuming process as compared to that of the reaction condition B (200 RPM agitation speed and 0.5 l/min CO_2 injection speed). Therefore, the reaction condition B is considered to be the most suitable for the carbonation of slaked lime, which is used for the production of the carbonated lime in this investigation.

Table 3. Moles of the reactants and products before and after the carbonization reaction and the summary of the CO_2 consumption in four different reaction conditions.

Chemical constituents	No. of Exp	Moles of the reactants and products before and after the carbonization reaction in four different reaction conditions							
		Condition A		Condition B		Condition C		Condition D	
		Before	After	Before	After	Before	After	Before	After
Moles of $Ca(OH)_2$	(i)	0.0117	–	0.0117	–	0.0117	–	0.0117	–
	(ii)	0.0117	–	0.0117	–	0.0117	–	0.0117	–
	(iii)	0.0117	–	0.0117	–	0.0117	–	0.0117	–
Moles of CO_2	(i)	0.0110	–	0.0127	–	0.0096	–	0.0122	–
	(ii)	0.0103	–	0.0112	–	0.0103	–	0.0101	–
	(iii)	0.0107	–	0.0115	–	0.0101	–	0.0114	–
Average		0.0107 ± 0.0004	–	0.0118 ± 0.0008	–	0.0100 ± 0.0004	–	0.0112 ± 0.0011	–
Moles of $CaCO_3$	(i)	–	0.0095	–	0.0118	–	0.0097	–	0.0113
	(ii)	–	0.0094	–	0.0116	–	0.0098	–	0.0109
	(iii)	–	0.0098	–	0.0113	–	0.0098	–	0.0110
Average		–	0.0096 ± 0.0002	–	0.0116 ± 0.0003	–	0.0098 ± 0.000	–	0.0111 ± 0.0002
Mole % of CO_2 consumption[a]		91.2 ± 3.0		100.1 ± 6.8		85.5 ± 3.1		96.0 ± 9.1	

[a]Mole percent of CO_2 consumed with respect to the mole of lime ($Ca(OH)_2$).

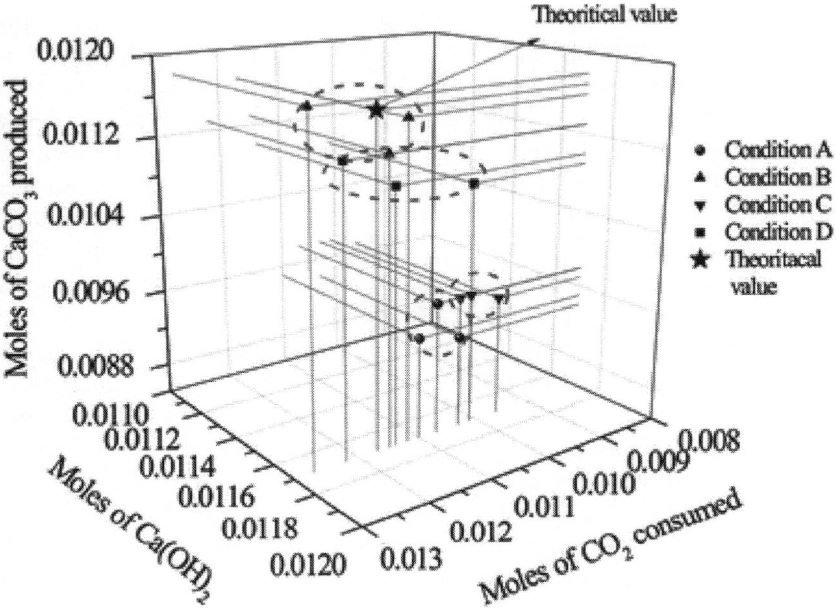

Figure 4. Moles of slaked lime (calcium hydroxide) and CO_2 used and the moles of calcium carbonate produced in the theoretical carbonation reaction scheme and the four different carbonation reaction conditions such as *A*, *B*, *C*, and *D*.

3.3 Thermo Gravimetric Analysis (TGA) of the Carbonated Material

The thermogravimetric analysis of the precipitates obtained from the aqueous carbonation reaction was performed to estimate the extent of carbonated product. In this investigation, the aqueous carbonation reaction in four different conditions was repeated three times to identify the most effective reaction condition. Figure 5 presents the TG thermogram (one replicate plot from each reaction condition) of the carbonated product obtained from the four different reaction conditions. As presented in the figure, the mass losses in the temperature range 400–550 and 600–800 °C are observed to be due to the decomposition of the calcium hydroxide and calcium carbonate, respectively (Chakraborty et al. 2013). From the figure, it is envisaged that irrespective of the reaction conditions used for the carbonation of slaked lime, the thermogram of the carbonated products shows a prominent mass loss in the temperature range 600–830 °C. However, the products obtained from the different reaction conditions show the different mass losses in the same temperature range. From the figure, the average mass loss attributed to be due to the decomposition of the calcium carbonate in four different reaction conditions, i.e., A, B, C and D is measured to be 29.4 ± 2.3, 39.2 ± 2.5, 31.9 ± 1.8 and 35.1 ± 2.3 %, respectively. It indicates that the different carbonation reaction conditions produce different amounts of carbonated product. Additionally, it is observed that the reaction condition B produces a greater extent of calcium carbonate as compared to that of the reaction conditions, i.e., A, C and D. Hence, the reaction condition B is considered to be most appropriate for the aqueous carbonation reaction. Additionally, it is apparent that the aqueous carbonation reaction successfully consumes the CO_2 to produce a solid residue (calcium carbonate). This can be further clarified by the X-ray diffraction analysis as well as FESEM analysis.

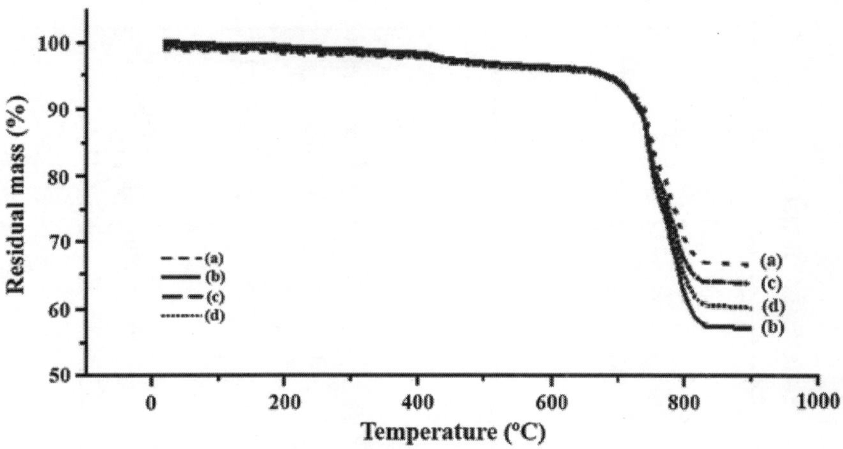

Figure 5. TG thermograms of the products obtained from the aqueous carbonation reaction in four different conditions. In the figure *a*, *b*, *c*, and *d* indicate the TG thermogram of the carbonated product obtained from the carbonation reaction condition *A*, *B*, *C*, and *D*, respectively.

3.4 X-ray Diffraction Analysis of the Aqueous Carbonated Product

The X-ray diffraction analysis of the aqueous carbonated lime was performed to identify the chemical phases present in the reaction product. Figure 6 depicts the X-ray diffraction pattern of the products obtained from the carbonation reaction in four different conditions. As presented in the figure, irrespective of the reaction condition used for the aqueous carbonation, the X-ray diffractogram of all samples possesses identical peaks. In the X-ray diffraction pattern, the characteristic peaks at the 2θ equal to 23.02°, 29.41°, 35.97°, 39.4, 43.15, 47.49 and 48.51°, 57.4° appear to be due to the calcium carbonate (calcite) (Stutzman 1996; Jiao et al. 2009; Kontoyannis and Vagenas2000; Han et al. 2011). Additionally, the very small peak at the 2θ equal to 56.42° and 57.94° appears to be due to the calcium carbonate also (Crowley 2010). Excluding the calcium carbonate peak, no other peak has been observed in the X-ray diffraction pattern. In fact, no peak for the calcium hydroxide (reactant) at the 2θ equal to 18.09°, 28.66°, 34.09°, 47.12°, 50.79° and 54.34° (Stutzman 1996) has been observed. Hence, from the X-ray diffraction analysis, it is demonstrated that the aqueous carbonation reaction is considered to be very effective in producing the calcium carbonate from the slaked lime. Additionally, it is assumed that the aqueous carbonation of the slaked lime could be an efficient method to consume the environmental CO_2.

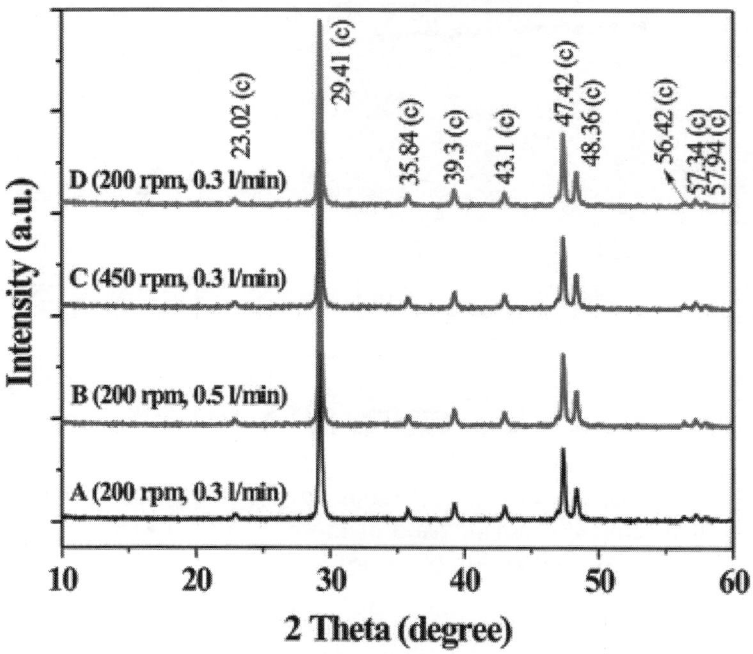

Figure 6. X-ray diffraction patterns of the products obtained by the aqueous carbonation reaction in four different conditions.

3.5 FESEM Analysis

Figure 7 shows the microstructure of the precipitates obtained from the aqueous carbonation reaction in four different conditions. As presented in the figure, the images (Figs. 7a to 7d) obtained from the FESEM analysis are observed to be similar. Hence, it is confirmed that the reaction condition does not affect the microstructure of the carbonated product. Additionally, the microstructure of the carbonated compound obtained from the four different reaction condition is observed to be a crystalline. Generally, the calcium carbonate possesses three polymorphs such as calcite (rhombohedral crystal structure), vaterite (globe or oval shaped) and aragonite (needle-shaped crystal structure) (Han et al. 2011; Stutzman 1996). As evidenced by the XRD analysis, irrespective of the reaction condition, the aqueous carbonation of slaked lime produces only calcite. Additionally, no crystal of the oval shaped and needle-shaped has been observed from the FESEM image of the carbonated product. Therefore, the crystal structure of the carbonated product (calcite) obtained from the aqueous carbonation reaction is considered to be rhombohedral.

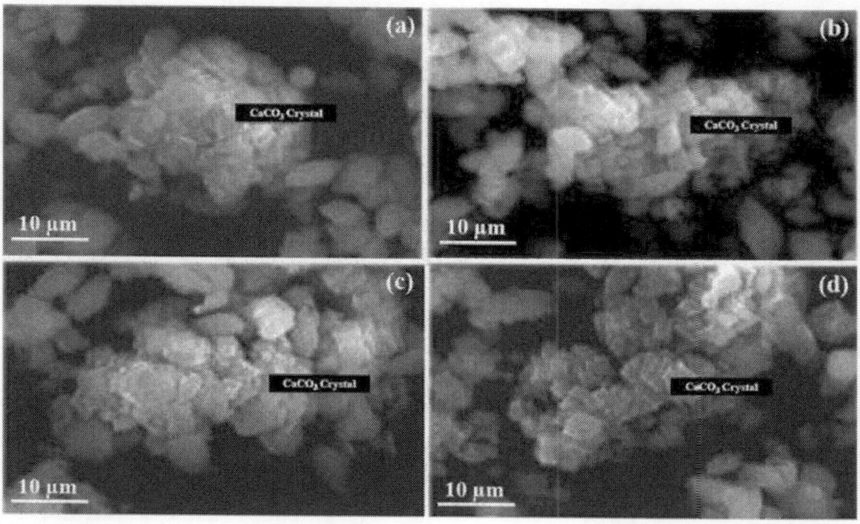

Figure 7. FESEM images of the products obtained from the carbonation reaction of the slaked lime in four different conditions. **a** Product obtained from the reaction condition A, **b** Product obtained from the reaction condition B, **c** Product obtained from the reaction condition C, **d** Product obtained from the reaction condition D.

Figures 8a and 8b depict the FESEM images of the commercially available ordinary Portland cement and the cementitious material produced by the hydrothermal method, respectively. From the Figure 8a, the particle of the ordinary Portland cement is observed to be irregular shaped. Whilst, the particles of the synthesized cementitious material are observed to be spherical shaped (Figure 8b). Additionally, an excellent particle size distribution of the synthesized cementitious material has been observed. The TGA, XRD, and FESEM analysis demonstrate that the aqueous carbonation of slaked lime produces rhombohedral crystal of calcite. However, the microstructure of the alternative cementitious material does not contain any rhombohedral crystal of calcite. Therefore, it is considered that the entire amount of calcium carbonate is consumed in producing the alternative cementitious material. Hence, it is assessed that the hydrothermal method can able to store the carbonated lime as a cementitious material.

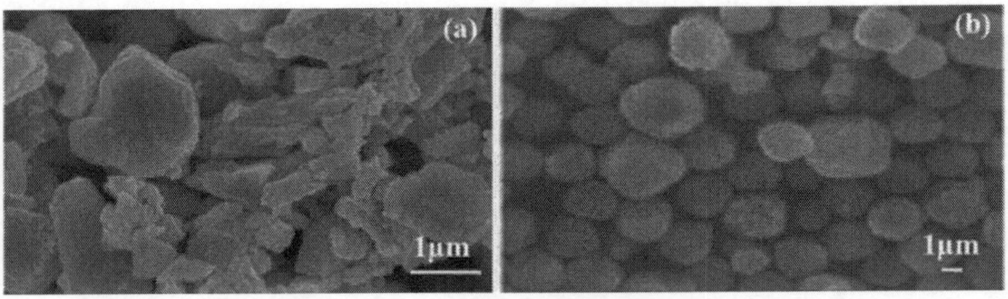

Figure 8. FESEM images of the **a** ordinary Portland cement (Jo et al. 2014c) and **b** the cementitious material produced by hydrothermal synthesis method utilizing aqueous carbonated material.

Figure 9 presents the microstructure of the 28 days cured control cement mortar (a) as well as the ACM based mortar (b). From Figure 9b, it is envisaged that the alternative cementitious material can able to produce a compact microstructure after hydration. Comparing the FESEM image of the ACM based mortar with the control mortar, the microstructure of the ACM based mortar is observed to be similar to that of the ordinary Portland cement based mortar. It is a well-known that the development of the microstructure is a crucial factor in enhancing the mechanical strength of the cementitious material. In this study, the cementitious material produced by the hydrothermal method utilizing carbonated lime develops a compact microstructure, which may contribute to improving the strength of the ACM based mortar.

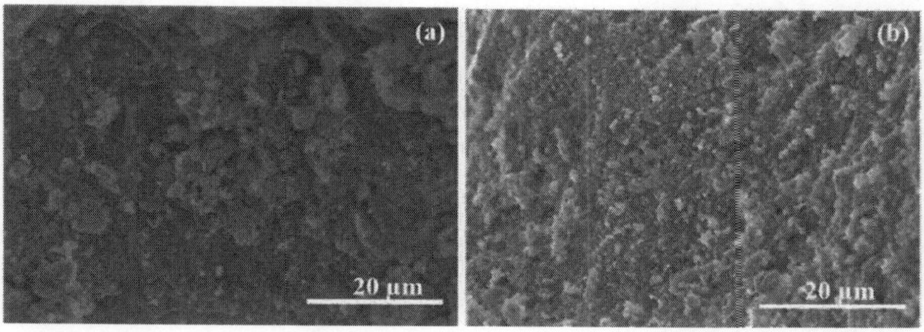

Figure 9. FESEM images of the 28 days hydrated **a** control cement as well as the **b** ACM based mortar sample.

3.6 Particle Size Analysis of the Alternative Cementitious Material

Figure 10 depicts the particle size distribution pattern of the alternative cementitious material. From the figure, it is envisaged that the particle size of the alternative cementitious material belongs in the range between 0.49–9.2 μm. The average particle size of the alternative cementitious material is measured to be 1.4 μm. Whilst, the particle size of the commercially available ordinary Portland cement is reported to be 10–30 μm (Jo et al. 2014c). Therefore, it is apparent that the particle size of the alternative cementitious material produced by the hydrothermal method is significantly smaller as compared to that of the commercially available Portland cement. Additionally, it is ascertained that the hydrothermal synthesis method is able to produce a spherical shaped submicron particle. Additionally, the specific gravity and fineness modulus of the alternative cementitious material produced by the hydrothermal method are measured to be 2.23 and 68,500 cm^2/g, respectively. Whilst, these are reported to be 3.15 and 2800 cm^2/g, respectively, for ordinary Portland cement (Jo et al. 2014c). As the alternative cementitious material possesses a smaller particle size and greater fineness modulus as compared to that of the OPC, therefore, it is assumed that the alternative cementitious material would expose a greater surface area for the hydration reaction, which may lead to developing the early strength.

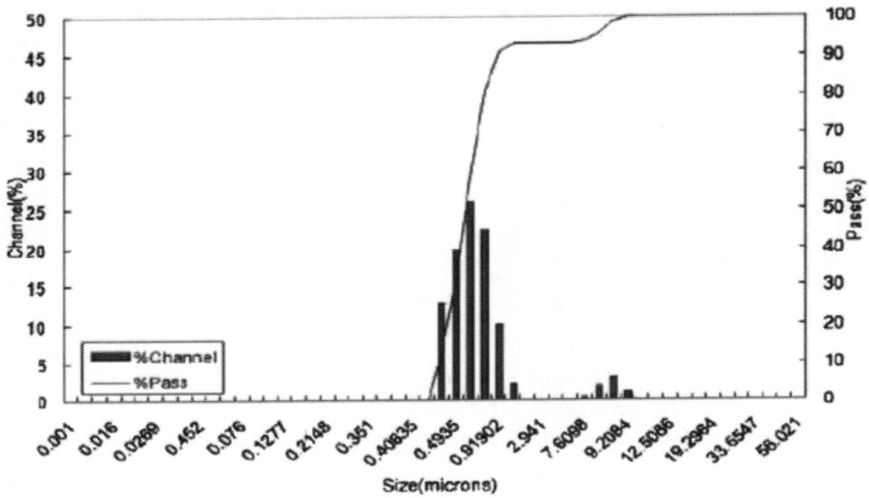

Figure 10. The particle size distribution curve of the alternative cementitious material produced by the hydrothermal method.

3.7 Chemical Composition Analysis

The oxide composition of the alternative cementitious was measured to evaluate the chemical characteristic. The oxide compositions of the alternative cementitious material are observed to be similar to that of the ordinary Portland cement. However, the amounts of the oxide phases are observed to be different. The mass (%) of the oxide phases such as SiO_2, CaO, Al_2O_3, Na_2O, Fe_2O_3 of alternative cementitious material are measured to be 50.84, 9.69, 28.97, 2 and 1.90 %, respectively. Whilst, the same are reported be 20.36, 64.33, 5.77, 0 and 2.84 %, respectively, for ordinary Portland cement (Jo et al. 2014a). Analyzing the oxide composition, it is envisaged that the alternative cementitious material contains a greater amount of SiO_2 and Al_2O_3 as compared to that of the ordinary Portland cement. Therefore, it is expected that the alternative cementitious material may produce a geopolymer compound after hydration reaction (Mayer et al. 2013; Kar et al. 2014). Additionally, it is assumed that the geopolymer material would produce an interpenetrating net structure in the bulk of the mortar, which may lead to developing the strength of the ACM based mortar.

3.8 Setting Time Analysis

Setting time is an important characteristic of the of the hydraulic cement. Usually, the setting time of the cement is measured to know the time required to set and harden the cement sample. In this investigation, the setting time is measured to know the time required for setting and hardening of the alternative cementitious material. The initial and final setting time of the alternative cementitious material is measured to be 242 ± 6 and 528 ± 5 min, respectively. Whereas, the initial and final setting time of the ordinary Portland cement is measured to be 210 ± 5 and 300 ± 8 min, respectively (Jo et al. 2014a). It is reported elsewhere that the initial and final setting time of a standard hydraulic cement should be more than 60 min and less than 10 h, respectively, at the ambient condition (KS L 5201 2013). In this study, the cementitious material produced by the hydrothermal method possesses a similar characteristic. Hence, it is assumed that the alternative cementitious material produced by the hydrothermal method can be considered as an alternative binder for the fabrication of mortar and concrete.

3.9 Compressive Strength

In this investigation, the mortar samples were fabricated using the ACM, sand and varying amounts of water and alkali activator. Table 2 depicts the mix proportion and sample code of the different types of the mortar samples. In this study, six samples from each batch were tested to obtain an average result of the compressive strength. Figure 11 shows the compressive strength of the control as well as ACM based mortar. As presented in the figure, the compressive strength of control mortar (CCM), ACM-M1 (ACM denotes alternative cementitious material and M indicates mortar), ACM-M2, and ACM-M3 is measured to be 34.2 ± 1.2, 25.4 ± 1.5, 28.6 ± 1.2, 33.8 ± 1.3 MPa, respectively. Additionally, it is envisaged that the compressive strength of the ACM based mortar increases gradually with the increase in alkali activator content. Interestingly, the compressive strength of the ACM based mortar fabricated using the maximum amount of alkali activator (ACM-M3) is observed to be comparable with the control mortar. The lower compressive strength of the mortar prepared using less extent of alkali activator is observed to be due to the inadequate hydration of the alternative cementitious material. In fact, due to the lack of driving force to transpire the hydration of the alternative cementitious material in the presence of water (Jo et al.2014a), the ACM demands high dose of alkali activator. The high dose of alkali activator supplies the driving force for the hydration of the alternative cementitious material, which in turn develops a compact microstructure. Eventually, the strength of the mortar increases. Hence, it is demonstrated that the ACM based mortar possesses a strong ability to gain strength in the presence of alkali activator. Accordingly, it is apparent that the alternative cementitious material could be used as a primary binder for the fabrication of mortar and concrete.

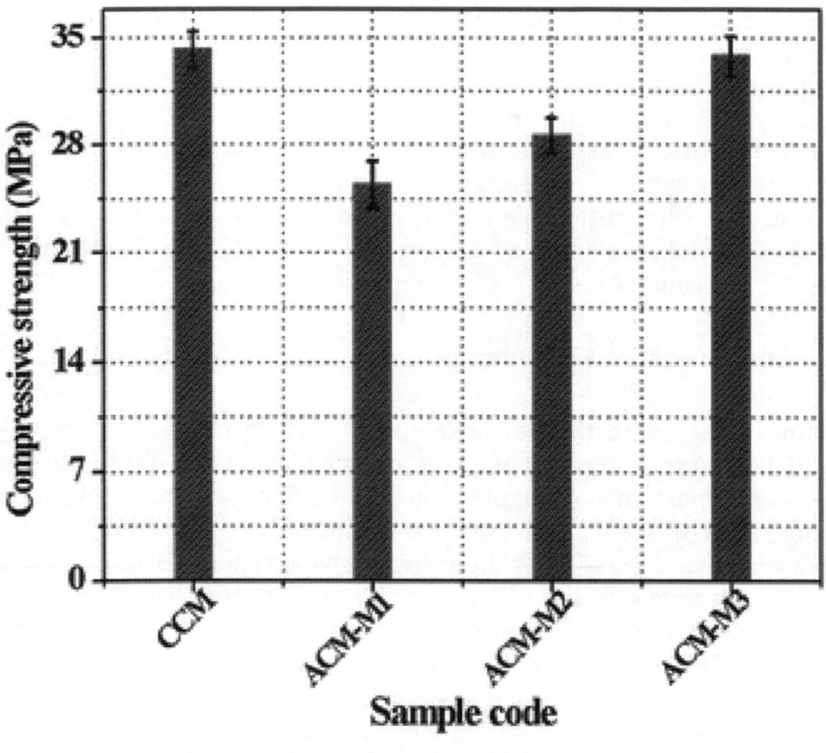

Figure 11. Compressive strength of the 28 days cured control cement mortar (CCM) and alternative cement based mortar (ACM-M1, ACM-M2, ACM-M3).

4 DISCUSSION

Viewing in light of the results presented in the preceding sections, it is apparent that the aqueous carbonation of slaked lime can accumulate the CO_2. Additionally, the hydrothermal method successfully produces a cementitious material utilizing the aqueous carbonated lime. In this section, we have tried to establish a reaction scheme for the aqueous the carbonation of the slaked lime and a plausible model for the hydrothermal synthesis of the cementitious material. The Eq. (2) represents a plausible reaction scheme for the carbonation of the slaked lime. Figure 12 depicts the plausible schematic view of the carbonation reaction steps. At the beginning of the carbonation reaction, the slaked lime solution contains the calcium (Ca^{2+}) and hydroxyl ion (OH^-) (Eq. (3)) (Olajire 2013), which in turn increases the pH of the solution (12.4). However, the purging of CO_2 into the reaction medium leads to decrease the pH of the reaction system by dissolving the CO_2 in the aqueous medium (Eq. 4). During this time, the aqueous CO_2 reacts with hydroxyl ions (OH^-) to form bicarbonate ions (Eq. (5)) However, the lifetime of this bicarbonate ion in the alkali medium is very small (Han et al.2011), which, therefore, reacts with the other hydroxyl ion to form a carbonate ion and water (Eq. (6)). In fact, in this step, the hydroxyl ions are consumed to neutralize the bicarbonate ions (Han et al. 2011), which in turn leads to reduce the pH of the medium. Additionally, in this step, due to the production of water, the total volume of water in the reaction system increases, which in turn reduces the OH^- concentration in the per unit volume of water. Eventually, the pH of the reaction medium decreases. Finally, the aqueous medium contains the calcium (Ca^{2+}) and carbonate ($CO_3{}^{2-}$) ions. The combination of these two ions forms the calcium carbonate as a white color precipitate (Eq. 7). The excess amount of CO_2 may present in the aqueous medium in the form of carbonic acid, which may lead to decrease the pH of the medium below 6 (Figure 12).

$$Ca(OH)_2 + CO_2 + H_2O \rightarrow CaCO_3 + 2H_2O \tag{2}$$

$$Ca(OH)_2\ (S) \rightleftharpoons Ca^{2+}\ (Aq) + 2OH^-\ (Aq) \tag{3}$$

$$CO_2\ (g) \rightleftharpoons CO_2\ (Aq) \quad (4)$$

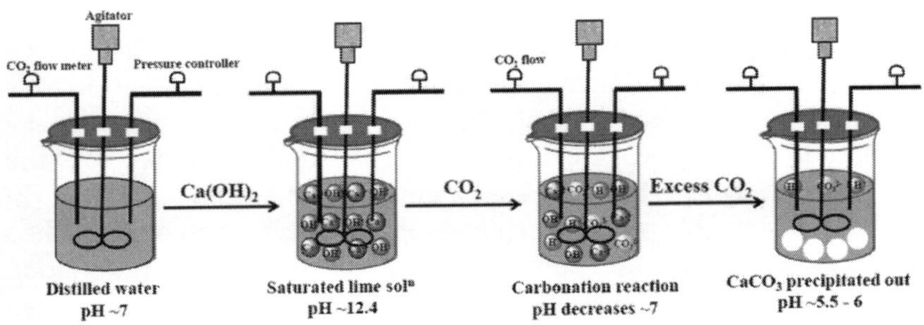

Figure 12. The plausible schematic view of the carbonation reaction steps.

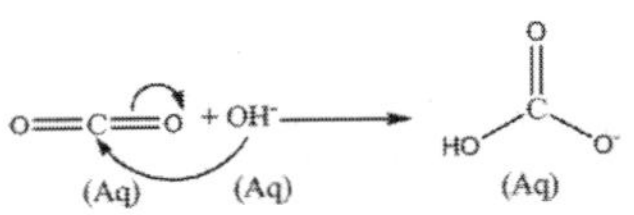

(5)

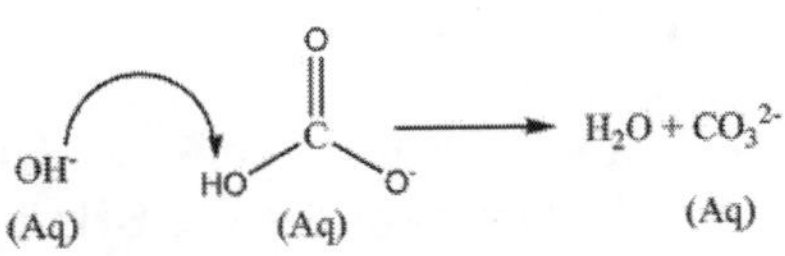

(6)

$$Ca^{2+}\ (Aq) + CO_3^{2-}\ (Aq) \longrightarrow CaCO_3\ (S)\downarrow \quad (7)$$

The carbonated material produced from the aqueous carbonation was used for the hydrothermal synthesis of the cementitious material. In this study, the alternative cementitious material was produced using aqueous carbonated lime infused with silica and hydrated alumina. Figure 2 represents a plausible model for the hydrothermal synthesis of the cementitious material. Xu and Deventer (2000) stated that a geopolymer compound can be produced from the silica infused with hydrated alumina. Hence, it is assumed that the alternative cementitious material produced in this investigation may contain geomaterial. During the hydrothermal synthesis, the mixture of the silica fume and calcium carbonate may produce a calcium bonded hydrated silica, (Eq. 8). Additionally, the hydrolysis of sodium aluminate may produce a hydrated alumina (Eq. (9)). After aging, the compounds produced by the hydrolysis (Eqs. (8)) and (9)) lead to produce the gel. Subsequently, the mixing of these gel compounds may produce the calcium bonded aluminosilicates (Eq. 10). As the alternative cementitious material contains calcium-substituted aluminosilicate geopolymer compound, therefore, the hydration of this cementitious material in the presence of the water is difficult. This could be due to the lack of driving force to initiate the hydration reaction of the alternative cementitious material (Jo et al. 2014a). Nevertheless, the use of alkaline solution (50 % NaOH) may supply the driving force to initiate the hydration reaction and lead to hardening the sample. During the hydration of the ACM in the presence of alkali activator, the mixture of geopolymer compounds may produce an interpenetrating net structure of the hydrated alumino-silicate compounds (Xu and Deventer 2000). Particularly, the hydration of the alternative cementitious material possibly produces the calcium-sodium aluminosilicate hydrate gel (Myers et al. 2013; Kar et al. 2014), which may hold the crystalline component and develop a compact microstructure. This phenomenon may, therefore, contribute to enhancing the strength of the ACM based mortar.

$$SiO_2 + CaCO_3 + H_2O \longrightarrow (HO)_3Si-O-Si(OH)_3 + CaCO_3 + \left[Si(OH)_2-O-Si(OH)_2 \right]_n -O-Ca- \left[Si(OH)_2-O-Si(OH)_2 \right]_n \quad (8)$$

$$NaAlO_2 + H_2O + NaOH \rightarrow NaAl^{-}(OH)_4 \quad (9)$$

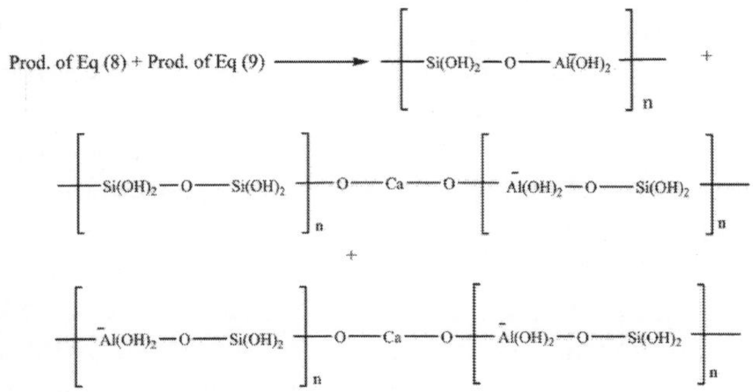

(10)

5 CONCLUSIONS

The paper presents a unique technique to produce an using the aqueous carbonated lime, silica fume, and hydrated alumina. Prior to producing the ACM, the aqueous carbonation of the slaked lime was performed in four different conditions to produce the carbonated lime. Analyzing the carbonated product using different analytical tools such as TGA, XRD, and FESEM, it is revealed that the carbonated product is calcium carbonate (calcite). Additionally, it is predicted that the carbonation reaction condition B (agitation speed 200 RPM and CO_2 injection speed 0.5 l/min) consumes a maximum amount carbon dioxide (0.0118 ± 0.0008 mol or 100.1 % with respect to moles of slaked lime) and produces a maximum amount of calcium carbonate (0.0116 ± 0.0003 mol). Besides, the alternative cementitious material was synthesized using the carbonated lime produced by the reaction condition B. Based on the particle size analysis, setting time measurement and compressive strength analysis, it is demonstrated that the ACM can control the physical and mechanical properties of the mortar. The compressive strength of the ACM based mortar (ACM-M3) prepared using the maximum amount of alkali activator is measured to be 33.8 ± 1.3 MPa, whilst the same is measured to be 34.2 ± 1.2 for the control mortar. Interestingly, it is revealed that the compressive strength of the ACM based mortar increases gradually with the increase in alkali activator content. Moreover, a plausible model has been proposed to explain the aqueous carbonation and the synthesis of the ACM. Hence, it is concluded that an alternative cementitious material can be produced by the hydrothermal method utilizing aqueous carbonated lime, which could be used as a primary binder for the fabrication of mortar. Accordingly, the utilization of the aqueous carbonated lime in producing a cementitious material is assumed to be an initial alternative approach to reduce the atmospheric CO_2.

ACKNOWLEDGMENTS

Author would like to acknowledge the BK21+, the Government of Korea (Republic of) for their funding to pursue this research program.

REFERENCES

1. Amato, I. (2013). Concrete solutions. *Nature News, 494*, 300–301.
2. Chakraborty, S., Kundu, S. P., Roy, A., Adhikari, B., & Majumder, S. B. (2013). Effect of jute as fiber reinforcement controlling the hydration characteristics of cement matrix. *Industrial and Engineering Chemistry Research, 52*, 1252–1260.
3. Chen, Z. Y., O'Connor, W. K., & Gerdemann, S. J. (2006). Chemistry of aqueous mineral carbonation for carbon sequestration and explanation of experimental results. *Environmental Progress & Sustainable Energy, 25*(2), 161–166.
4. Chindaprasirt, P., & Cao, T. (2015). Setting, segregation and bleeding of alkali-activated cement mortar and concrete binders. In F. P. Torgal, J. A. Labrincha, C. Leonelli, A. Palomo, & P. Chindaprasirt (Eds.), *Handbook of alkali-activated cements, mortars and concretes* (pp. 113–131)., Woodhead Publishing series in Civil and Structural Engineering, No.: 54 Cambridge: Woodhead publishing.
5. Chizmeshya, A. V. G., McKelvy, M. J., Marzke, R., Ito, N., Wolf, G., Bèarat, H., et al. (2007) Investigating geological sequestration reaction processes under in situ process conditions. 32nd International Technical Conference on Coal Utilization & Fuel Systems, 441.

6. Criado, M., Fernandez-Jimenez, A., & Palomo, A. (2010). Alkali activation of fly ash. Part III: effect of curing conditions on reaction and its graphical description.*Fuel, 89*, 3185–3192.

7. Crowley, S. F. (2010). Mineralogical and chemical composition of international carbon and oxygen isotope calibration material NBS 19, and reference materials NBS 18, IAEA-CO-1 and IAEA-CO-8. *Geostandards and Geoanalytical Research, 34*(2), 193–206.

8. Dinakar, P., Sahoo, P. K., & Sriram, G. (2013). Effect of metakaolin content on the properties of high strength concrete. *International Journal of Concrete Structures and Materials, 7*(3), 215–223.

9. Feely, R. A., Sabine, C. L., Lee, K., Berelson, W., Kleypas, J., Fabry, V. J., et al. (2004). Impact of anthropogenic CO_2 on the $CaCO_3$ system in the oceans. *Science,305*, 362–366.

10. Galan, I., Andrade, C., Mora, P., & Sanjuan, M. A. (2010). Sequestration of CO_2 by concrete carbonation. *Environmental Science and Technology, 44*, 3181–3186.

11. Gerdemann, S. J., O'Connor, W. K., Dahlin, D. C., Penner, L. R., & Rush, H. (2007). Ex-situ aqueous mineral carbonation. *Environmental Science and Technology,41*, 2587–2593.

12. Han, S. J., Yoo, M., Kim, D. W., & Wee, J. H. (2011). Carbon dioxide capture using calcium hydroxide aqueous solution as the absorbent. *Energy & Fuels, 25*, 3825–3834.

13. Hanchen, M., Prigiobbe, V., Baciocchi, R., & Mazzotti, M. (2008). Precipitation in the mg-carbonate system effects of temperature and CO_2 pressure. *Chemical Engineering Science, 63*, 1012–1028.

14. Huijgen, W. J. J., Comans, R. N. J. (2005). *Carbon dioxide sequestration by mineral carbonation*. Literature review update (2003–2004), ECN-C-05-022. Energy Research Centre of The Netherlands, Petten, Netherlands. Available at: http://www.ecn.nl/docs/ library/report/2005/ c05022.pdf. Accessed: 23 Dec 2014.

15. Huijgen, W. J. J., Comans, R. N. J., & Witkamp, G. J. (2007). Cost evaluation of CO_2 sequestration by aqueous mineral carbonation. *Energy Conversion and Management, 48*, 1923–1935.

16. Huijgen, W. J. J., Witcamp, G. J., & Comans, R. N. J. (2004). Mineral CO_2 sequestration in alkaline solid residues. *Proceedings Materials of 7th International Conference on Greenhouse Gas Control Technologies* (pp. 2415–2418) Vancouver, BC.

17. Huijgen, W. J. J., Witkamp, G. J., & Comans, R. N. J. (2006). Mechanisms of aqueous wollastonite carbonation as a possible CO_2 sequestration process. *Chemical Engineering Science, 61*, 4242–4251.

18. Jacobsen, J., Rodrigues, M. S., Telling, M. T. F., Beraldo, A. L., Santos, S. F., Aldridge, L. P., et al. (2013). Nano-scale hydrogen-bond network improves the durability of greener cements. *Scientific Reports, 3*(2667), 1–6.

19. Jeon, D., Jun, Y., Jeong, Y., & Oh, J. E. (2015). Microstructural and strength improvements through the use of Na_2CO_3 in a cementless $Ca(OH)_2$-activated Class F fly ash system. *Cement and Concrete Research, 67*, 215–225.

20. Jiao, J., Liu, X., Gao, W., Wang, C., Feng, H., Zhao, X., et al. (2009). Two-step synthesis flowerlike calcium carbonate/biopolymer composite materials.*CrystEngComm, 11*, 1886–1891.

21. Jo, B. W., Chakraborty, S., Jo, J. H., & Lee, Y. S. (2015). Effectiveness of carbonated lime as a raw material in producing a CO_2-stored cementitious material by the hydrothermal method. *Construction and Building Materials, 95*, 556–565.

22. Jo, B. W., Chakraborty, S., & Kim, K. H. (2014a). Investigation on the effectiveness of chemically synthesized nano cement in controlling the physical and mechanical performances of concrete. *Construction and Building Materials, 70*, 1–8.

23. Jo, B. W., Chakraborty, S., Kim, K. H., & Lee, Y. S. (2014b). Effectiveness of the top-down nanotechnology in the production of ultrafine cement (~220 nm).*Journal of Nanomaterials, 57*, 1–9.

24. Jo, B. W., Chakraborty, S., & Yoon, K. W. (2014c). Synthesis of a cementitious material nanocement using bottom-up nanotechnology concept: An alternative approach to avoid CO_2 Emission during production of cement. *Journal of Nanomaterials, 97*, 1–12.

25. Juenger, M., Winnefeld, F., Provis, J., & Ideker, J. (2011). Advances in alternative cementitious binders. *Cement and Concrete Research, 41*, 1232–1243.

26. Kar, A., Ray, I., Halabe, U. B., Unnikrishnan, A., & Dawson-Andoh, B. (2014). Characterizations and quantitative estimation of alkali-activated binder paste from microstructures. *International Journal of Concrete Structures and Materials, 8*(3), 213–228.

27. Keeling, C. D., Whorf, T. P., Wahlen, M., & VanderPlicht, J. (1995). Inter-annual extremes in the rate of rise of atmospheric carbon dioxide since 1980. *Nature, 75*, 666–670.

28. Kim, M. S., Jun, Y., Lee, C., & Oh, J. E. (2013). Use of CaO as an activator for producing a price competitive non-cement structural binder using ground granulated blast furnace slag. *Cement and Concrete Research, 54*, 208–214.

29. Kontoyannis, C. G., & Vagenas, N. V. (2000). Calcium carbonate phase analysis using XRD and FT-Raman spectroscopy. *Analyst., 125*, 251–255.

30. Kotwal, A. R., Kim, Y. J., Hu, J., & Sriraman, V. (2015). Characterization and early age physical properties of ambient cured geopolymer mortar based on class C fly ash. *International Journal of Concrete Structures and Materials, 9*(1), 35–43.

31. KSF 2405. (2010). *Testing method for compressive strength of concrete*. Seoul, Korea: Bureau of Korean standard (in Korean).

32. KSL 5108. (2007). *Testing method for setting time of hydraulic cement by vicat needle*. Seoul, Korea: Bureau of Korean standard (in Korean).

33. KSL 5201. (2013). *Portland cement*. Seoul, Korea: Bureau of Korean standard (in Korean).

34. Metz, B., Davidson, O., deConinck, H., Loos, M., & Meyer, L., (Eds.) (2005). *IPCC special report on carbon dioxide capture and storage*. Cambridge University Press, New York, NY: 431. Available at: http://www.ipcc.ch/pdf/ special-reports/srccs/srccs_wholereport.pdf. Accessed: 4 Nov 2014.

35. Myers, R. J., Bernal, S. A., Nicolas, R. S., & Provis, J. L. (2013). Generalized structural description of calcium–sodium

aluminosilicate hydrate gels: The cross-linked substituted tobermorite model. *Langmuir, 29,* 5294–5306.

36. Olajire, A. A. (2013). A review of mineral carbonation technology in sequestration of CO_2. *Journal of Petroleum Science and Engineering, 109,* 364–392.

37. Phair, J. (2006). Green chemistry for sustainable cement production and use. *Green Chemistry, 8,* 763–780.

38. Roychand, R., De Silva, S., Law, D., & Setunge, S. (2016). Micro and nano engineered high volume ultrafine fly ash cement composite with and without additives.*International Journal of Concrete Structures and Materials.*

39. Schrabback, J. M. (2010). Concepts for 'green' cement. ICR. Available at: www.sika.com/dms/get//Concepts%20for%20Green%20Cement.pdf. Accessed 11 March 2015.

40. Siegenthaler, U., & Oeschger, H. (1987). Biospheric CO2 emissions during the past 200 years reconstructed by deconvolution of ice core data. *Tellus, 39B,* 140–154.

41. Stutzman, P.E. (1996). Guide for X-ray powder diffraction analysis of portland cement clinker. NISTIR 5755. Building and fire research laboratory, National Institute of standards and Technology, U.S Department of Commerce, Gaitheresburg, MD. Available at: http://fire.nist.gov/bfrlpubs /build96/PDF/b96138.pdf. Accessed 4 Nov 2014.

42. Xu, Hua, & Van Deventer, J. S. J. (2000). The geopolymerisation of alumino-silicate minerals. *International Journal of Mineral Processing, 59,* 247–266.

4
CHAPTER

A Review on Structural Behavior, Design, and Application of Ultra-High-Performance Fiber-Reinforced Concrete

Doo-Yeol Yoo[1], Young-Soo Yoon[2]

[1]*Department of Architectural Engineering Hanyang University Seoul 133-791, Korea.*

[2]*School of Civil, Environmental and Architectural Engineering Korea University Seoul 136-713, Korea.*

ABSTRACT

An overall review of the structural behaviors of ultra-high-performance fiber-reinforced concrete (UHPFRC) elements subjected to various loading conditions needs to be conducted to prevent duplicate research and to promote its practical applications. Thus, in this study, the behavior of various UHPFRC structures under different loading conditions, such as flexure, shear, torsion, and high-rate loads (impacts and blasts), were synthetically reviewed. In addition, the bond performance between UHPFRC and reinforcements, which is fundamental information for the structural performance of reinforced concrete structures, was investigated. The most widely used international recommendations for structural design with UHPFRC throughout the world (AFGC-SETRA and JSCE) were specifically introduced in terms of material models and flexural and shear design.

Lastly, examples of practical applications of UHPFRC for both architectural and civil structures were examined.

Keywords: ultra-high-performance fiber-reinforced concrete; bond performance; structural behavior; design code; application

1 INTRODUCTION

Ultra-high-performance fiber-reinforced concrete (UHPFRC), which was developed in the mid-1990s, has attracted much attention from researchers and engineers for practical applications in architectural and civil structures, because of its excellent mechanical performance, i.e., compressive strength is greater than 150 MPa and a design value of tensile strength is 8 MPa (AFGC-SETRA 2002), durability, energy absorption capacity, and fatigue performance (Farhat et al. 2007; Graybeal and Tanesi 2007; Yoo et al. 2014c; Li and Liu 2016). In particular, its very high strength properties result in a significant decrease in the structural weight, i.e., the weight of UHPFRC structures is about 1/3 (or 1/2) of the weight of general reinforced concrete (RC) structures at identical external loads (Tam et al. 2012). Consequently, slender structures, which are applicable for long-span bridges, can be fabricated with UHPFRC, leading to low overall construction costs.

In order to apply such a newly developed innovative material to real structures, numerous studies have been carried out in many countries in Europe, North America, and Asia. Since UHPFRC was first developed by France's research group, the first technical recommendation on UHPFRC for both material properties and structural design was introduced in France in 2002 and was called the AFGC-SETRA recommendation (AFGC-SETRA 2002). Thereafter, a state-of-the-art report on UHPFRC covering all material and design aspects was published in Germany in 2003 (DAfStB 2003). Then, in 2004, the Japan Society of Civil Engineers (JSCE) published their own design recommendations for UHPFRC based on Ductal® (Orange et al. 1999), commercial UHPFRC available in the world, (JSCE 2004). Lastly, in recent years, the Korea Concrete Institute (KCI) also developed a design code for UHPFRC (KCI 2012), similar to those in France and Japan, by using K-UHPC, another UHPFRC material developed by Korea Institute of Civil Engineering and Building Technology (Kim et al. 2008).

Due to the superb fiber bridging capacities of UHPFRC at cracked surfaces, leading to a special strain-hardening (or deflection-hardening) response with multiple micro-cracks, many researchers have focused on using it to the structures dominated by flexure, shear, and torsion. Furthermore, UHPFRC has also been considered as one of the promising materials for impact- and blast-resistant structures, because of its enhanced strength and energy absorption capacity, along with strain-hardening cementitious composites containing polymeric fibers (Astarlioglu and Krauthammer 2014; Choi et al. 2014). These properties can help to overcome the brittle failure of plain concrete, which has poor energy absorption capacity for impacts and blasts. Since the structural behaviors of UHPFRC under flexure, shear, and torsion, and when subjected to high-rate loadings, such as impacts and blasts, are highly sensitive to numerous factors, i.e., structural shape, loading condition, strain-rate, casting method, reinforcement ratio, etc., it is necessary to synthetically review the scattered studies.

The purpose of this research is to analyze the current state of knowledge of the structural behavior, design techniques, and applications of UHPFRC under various loading conditions As explained above, the attention of this paper is focused on (1) bond performance between UHPFRC and various reinforcements, which is basic information needed for the design of reinforced structures, (2) structural behavior of UHPFRC under flexure, shear, torsion, and high-rate loading, (3) the most widely used UHPFRC design recommendations in the world, and (4) examples of practical applications in both architectural and civil structures.

2 HISTORICAL DEVELOPMENT OF UHPFRC

Roy et al. (1972) and Yudenfreund et al. (1972) first introduced ultra-high-strength cementitious paste with low porosity in the early 1970s. With special curing methods using heat (250 °C) and pressure (50 MPa), Roy et al. (1972) achieved a cementitious paste with almost zero porosity and a compressive strength of approximately 510 MPa. On the other hand, Yudenfreund et al. (1972) obtained a cement paste having a compressive strength of about 240 MPa with normal curing (25 °C) for 180 days. To do this, Yudenfreund et al. (1972) provided a special treatment on-ground clinker, used the low water-to-cement ratio of 0.2, and Blaine surface areas ranging from 6000 to 9000 cm^2/g. After nearly 10 years, Birchall et al. (1981) and Bache (1981) could develop two types of ultra-high-strength paste (or concrete) with very low porosity, such as densified with small particles (DSPs) concrete and macro-defect free (MDF) paste, by developing a pozzolanic admixture and a high-range water-reducing agent. Birchall et al. (1981) achieved the development of the cement pastes with compressive strength over 200 MPa and flexural strengths of 60–70 MPa, by removing macroscopic flaws during material preparation without using fibers or high-pressure compaction. Bache (1981) also successfully developed the concrete that was DSPs and had a compressive strength of 120–270 MPa. The key technique to densely pack the spaces between the cement particles was to use ultra-fine particles and an extremely low water content, with a large quantity of high-range water-reducing agent. In the mid-1990s, Richard and Cheyrezy (1995) first introduced the concept of and mixing sequence for reactive powder concrete (RPC), which was the forerunner of UHPFRC. To obtain a very high strength, the granular size was optimized by the packing density theory, by excluding coarse aggregate and by providing heat (90 and 400 °C) and pressure treatments. In addition, 1.5–3 % (by volume) of straight steel microfibers, with a diameter of 0.15 mm and a length of 13 mm, were added to achieve high ductility; consequently, the RPC developed by Richard and Cheyrezy (1995) showed compressive strengths of 200–800 MPa and fracture energies up to 40 kJ/m^2.

3 PERFORMANCE OF STRUCTURAL UHPFRC ELEMENTS

3.1 Bond Behavior Between UHPFRC and Reinforcements

In order to practically apply a newly developed UHPFRC in the structures, bond performance with reinforcements should be examined. Many researchers (Jungwirth and Muttoni 2004; Ahmad Firas et al. 2011; Yoo et al. 2014a, b, 2015a) have investigated the bond behavior of internal steel and fiber-reinforced polymer (FRP) reinforcements with UHPFRC. Jungwirth and Muttoni (2004) carried out pullout test of deformed steel reinforcing bar using a 160 mm cube. Various bond lengths ranging from 20 to 50 mm and two different bar diameters of 12 and 20 mm were adopted. In their study, the average bond strength of steel bars embedded in UHPFRC was found to be 59 MPa, approximately 10 times higher than the bond strength of steel bars embedded in ordinary concrete, and the theoretical development length of deformed steel bars in UHPFRC was suggested by $l_b = f_y d_b / 4\tau_{max}$, where f_y is the yield strength of steel bar, d_b is the nominal diameter of steel bar, and τ_{max} is the bond strength. Yoo et al. (2014c) examined the effects of fiber content and embedment length on the bond behavior of deformed steel bars embedded in UHPFRC. For this, Yoo et al. (2014c) performed a number of pullout tests by modifying the test method, proposed by RILEM recommendations (RILEM 1994); the 150 mm cubic specimens with a single bar embedded vertically along the central axis were fabricated and used for testing. The embedment lengths were determined by 1 and 2 times the bar diameter, instead of using $5d_b$, as suggested by the RILEM recommendations. The bond strength was insignificantly affected by the fiber content and embedment length, but it clearly correlated with the compressive strength. The CEB-FIP Model Code (MC90) (CEB-FIP 1993), which defined τ_{max} as $2.0 f_c'^{0.5}$, substantially underestimated the bond strength of steel bars in UHPFRC because the parameters were suggested based on test data from previous concretes. Thus, Yoo et al. (2014c) proposed modified coefficients for the bond strength of steel bars in UHPFRC, based on a number of test data, as follows (Figure 1):

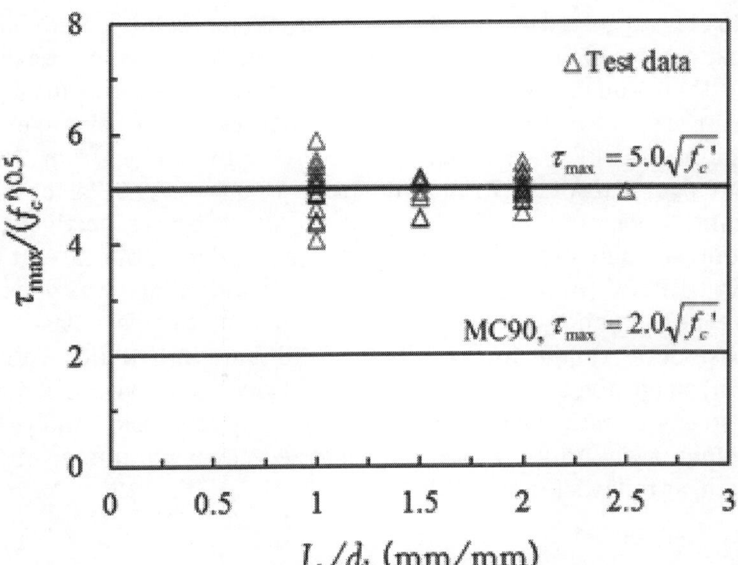

Figure 1. Normalized bond strengths of deformed steel bars embedded in UHPFRC.

$$\tau_{max} = 5.0\sqrt{f_c'} \tag{1}$$

In addition, CMR model (Cosenza et al. 1995), which sets $\tau = \tau_{max} \times (1 - e^{-s/sr})^{\beta}$, was found to be appropriate for simulating the ascending bond stress versus slip behavior of steel bars embedded in UHPFRC, and the parameters were proposed as $\tau_{max} = 5.0f_c'^{0.5}$, $s_r = 0.07$, and $\beta = 0.8$, where s_r and β are coefficients based on the curve fitting of test data.

Ahmad Firas et al. (2011) experimentally investigated the bond performance between carbon-fiber-reinforced polymer (CFRP) bars and UHPFRC according to the surface treatment, embedment length, bar diameter, and concrete age. Based on the test data, it was noted that the bond strength was insignificantly affected by the surface treatment of the glass-fiber-reinforced polymer (GFRP) bar; similar bond strengths for smooth bars and sand-coated bars were obtained. On the other hand, a decrease in bond strength was obtained by increasing both bar diameter and embedment length. The ultimate bond strength of CFRP bars in UHPFRC was insignificantly changed by age after 3 days, because it was primarily affected by the shear strength of the connection between the core and the outer layer of the CFRP bars. Ahmad Firas et al. (2011) suggested a development length for sand-coated bars of approximately $40d_b$, and a development length for a smooth bar of longer than $40d_b$. Yoo et al. (2015b) also examined the local bond behavior of GFRP bars embedded in UHPFRC. The average bond strengths of GFRP bars in UHPFRC were found to be from 16.7 to 22.8 MPa for a d_b of 12.7 mm, and from 19.3 to 27.5 MPa for a d_b of 15.9 mm, which are approximately 73 and 66 % less, respectively, than the bond strengths of deformed steel bars. Similar to the case of CFRP bars in UHPFRC (Ahmad Firas et al. 2011), bond failure was generated by the delamination of the resin and fiber in the bar. Based on a database of 68 pullout test results for GFRP bars in UHPFRC, Yoo et al. (2015b) suggested an equation for the relationship between normalized bond strength and development length by using regression analysis and by assuming no influence of the normalized cover parameter on bond strength, as follows (Figure 2b):

$$\frac{u}{\sqrt{f_c'}} = 1.05 + 0.85\frac{d_b}{L_e} \tag{2}$$

where u is the bond strength ($=\tau_{max}$), d_b is the bar diameter, and L_e is the embedment length.

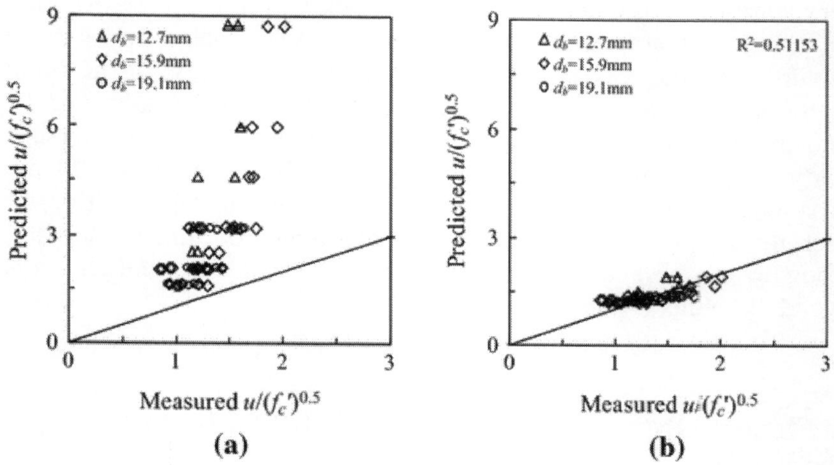

Figure 2. Comparison of measured and predicted values of normalized bond strength of GFRP bars in UHPFRC; **a** ACI 440.1R, **b** proposed equation (Yoo et al. 2015b).

The American Concrete Institute (ACI) 440.1R model (ACI 2006) was inappropriate for UHPFRC; it significantly overestimated the test data (normalized bond strength), as shown in Figure 2a.

Yoo et al. (2015b) also pointed out that the previous model for development length of FRP bar in concrete, suggested by Wambeke and Shield (2006), was not appropriate for UHPFRC; thus, they proposed an expression for the development length of GFRP bars in UHPFRC, which is only valid for the case of pullout failure, as follows:

$$L_{d,pullout} = \frac{d_b f_{fu}}{3.4\sqrt{f_c'}} \quad (3)$$

where $L_{d,pullout}$ is the development length and f_{fu} is the ultimate strength of rebar.

Schäfers and Seim (2011) performed experimental and numerical investigations on the bond performance between timber and UHPFRC. The glued-laminated timber was bonded to sandblasted and ground UHPFRC with the "Sikadur 330" epoxy resin. Regardless of the bond length and surface treatment, most of specimens showed failure of the bond in the timber close to the bond-line. Based on the Volkersens theory, Schäfers and Seim (2011) suggested a bond length of 400 mm for standard test method to evaluate the bond strength of timber-concrete composites and noted that the effect of tensile stresses, orthogonal to the bond-line, can be neglected when the bond length is beyond 300 mm.

3.2 Flexural Dominated Reinforced UHPFRC Beams, Girders, and Composite Structures

Due to its excellent post-cracking tensile performance with multiple micro-cracks occurred, UHPFRC has attracted attention from engineers for application in structural elements subjected to bending. Several international recommendations (AFGC-SETRA, JSCE, and KCI) from France, Japan, and South Korea (AFGC-SETRA 2002; JSCE2004; KCI 2012) thus provide stress–strain models for compressive and tensile stress blocks in the cross-section, as well as the detailed process of predicting the ultimate capacity of UHPFRC elements under flexure. Since strain (and stress) distribution in the cross-section varies according to the curvature of a beam, multilayer sectional analysis (Yoo and Yoon 2015) is required to calculate an appropriate neutral axis depth and moment capacity at a certain curvature level.

Yoo and Yoon (2015) first reported test results of a number of reinforced UHPFRC beams to investigate the effects of steel fiber aspect ratio and type on flexural performance. Since a portion of the tensile stress after cracking was resisted by the steel fibers, low reinforcement ratios (percent) of 0.94 and 1.50 % were selected. In order to prevent brittle shear failure, stirrups were conservatively designed based on the specimens made of ultra-high-performance concrete (UHPC) without fibers. From the test results (Figure 3), the beams made by UHPFRC with 2 % by volume of steel fibers exhibited much higher post-cracking stiffness and ultimate load capacity, compared to those made by UHPC without fiber, called 'NF'. In addition, the use of long straight or twisted steel fibers (S19.5, S30, and T30) led to a higher ductility than the use of short straight steel fibers (S13), which are applied for commercial UHPFRC available in North America (Graybeal 2008), at the identical fiber volume fraction. However, it is very interesting to note that much lower ductility indices were obtained by including steel fibers. This is caused by the fact that due to the very high bond strength between UHPFRC and steel rebar and its crack localization behavior, the steel rebar ruptured at a relatively smaller mid-span deflection, as compared with UHPC beams without fiber. Thus, Yoo and Yoon (2015) concluded that the strain-hardening behavior of UHPFRC was unfavorable to the ductility of reinforced beams.

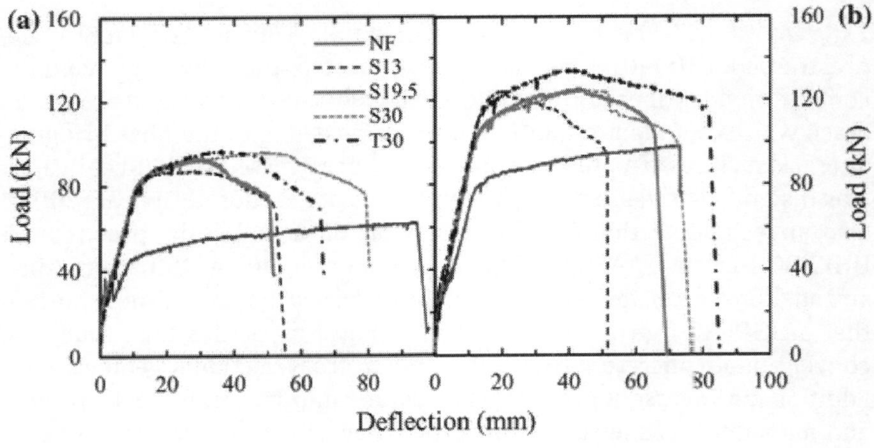

Figure 3. Load-deflection curves of steel bar-reinforced UHPFRC/UHPC beams; **a** ρ = 0.94 %, **b** ρ = 1.50 % [NF = UHPC w/o fiber, S13 = UHPFRC w/ straight steel fibers (L_f/d_f = 13/0.2 mm/mm), S19.5 = UHPFRC w/ straight steel fibers (L_f/d_f = 19.5/0.2 mm/mm), S30 = UHPFRC w/ straight steel fibers (L_f/d_f = 30/0.3 mm/mm), T30 = UHPFRC w/ twisted steel fibers (L_f/d_f = 30/0.3 mm/mm)] (Yoo and Yoon 2015).

In order to establish reasonable design codes for UHPFRC, Yang et al. (2010) carried out several four-point flexural tests for UHPFRC beams having reinforcement ratios less than 0.02. Test variables were the amount of steel rebar and the placement method. From their test results, placing concrete at the ends of the beams yielded better performance than when concrete was placed at the mid-length because of better fiber orientation to the direction of beam length at the maximum moment zone. In addition, they reported that all test beams showed a ductile response with the ductility index ranging from 1.60 to 3.75 and were effective in controlling cracks. However, the meaning of 'ductile response' could be incorrectly delivered to readers because no test results of reinforced UHPC beams without fiber were reported. In accordance with the test results by Yoo and Yoon (2015), reinforced UHPFRC beams exhibited lower ductility indices compared to beams without fiber due to the crack localization behavior, and Dancygier and Berkover (2016) also reported that the inclusion of steel fibers resulted in a decrease of flexural ductility of beams with low conventional reinforcement ratios.

Yang et al. (2011) examined the flexural behavior of large-scale prestressed UHPFRC I-beams. They indicated that the high volume content of steel fibers in UHPFRC effectively controlled the increase in crack widths, and led to multiple micro-cracks due to the fiber bridging at crack surfaces. The flexural strength of prestressed UHPFRC I-beams was insignificantly affected by the presence of stirrups. Graybeal (2008) also investigated the flexural behavior of a full-scale prestressed UHPFRC I-girder (AASHTO Type II girder) containing 26 prestressing strands. Based on the experimentally observed behavior, he reported that a UHPFRC I-girder shows larger flexural capacities than that of a conventional concrete girder with similar cross-sectional geometry. In addition, an inversely proportional relationship between crack spacing and maximum tensile strain was experimentally observed, as shown in Figure 4, and the following equation was suggested:

$$\varepsilon = 450 + \frac{2520}{\sqrt{s_{cr}}} + \frac{25{,}800}{\sqrt{s_{cr}^2}} \quad \text{(in mm)} \tag{4}$$

where ε is the tensile strain and s_{cr} is the crack spacing.

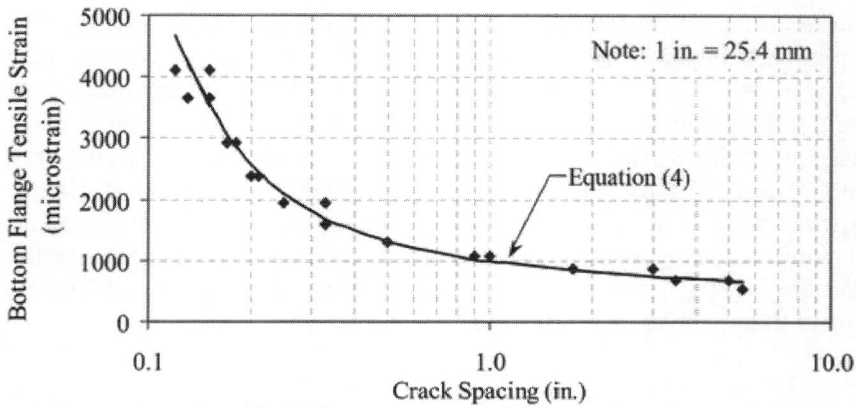

Figure 4. Tensile strain related to flexural crack spacing of UHPFRC I-girder (Graybeal 2008).

In recent years, several studies (Ferrier et al. 2015; Yoo et al. 2016) have been carried out to develop a new type of high-performance lightweight beams by applying UHPFRC and FRP rebar. Ferrier et al. (2015) investigated the structural behavior of I-shaped UHPFRC beams reinforced with CFRP and GFRP rebar, according to the rebar axial stiffness ranging from 9 MN to 30 MN. Experimental results indicated that the CFRP rebar was effective in increasing the bending stiffness, which results in a lower mid-span deflection, as compared with the case of the GFRP rebar due to the higher elastic modulus of the former. Thus, they concluded that the axial stiffness of the FRP reinforcement is the most influential parameter of bending stiffness of beams. Yoo et al. (2016) also examined the flexural behavior of UHPFRC beams reinforced with GFRP rebar and hybrid reinforcements (steel + GFRP rebar), according to the axial stiffness ranging from 13 to 95.5 MN. Hybrid reinforcements were considered in their study because it has been considered as one of the most promising methods to overcome the large service deflection problems of conventional FRP-reinforced concrete beams reported by several researchers (Lau and Pam 2010; Yoon et al. 2011). Due to the strain-hardening characteristics of UHPFRC, all tested beams provided very stiff load versus deflection response even after the formation of cracks (Figure 5), which is distinctive response with conventional FRP-reinforced concrete beams, and satisfied the service crack width criteria of the Canadian Standards Association (CAN/CSA) S806 (CAN 2002). Furthermore, the deformability factors suggested by Jaeger et al. (1995) were higher than the lower limit ($D_f = 4$) of CAN/CSA-S6 (CAN 2006) for all test beams.

Therefore, it was noted that the use of UHPFRC could be a new solution for solving the major drawbacks limiting the practical application of FRP rebar instead of steel rebar. An increase in the GFRP reinforcement ratio led to an improvement in the flexural performance, such as higher post-cracking stiffness, load carrying capacity, and ductility. However, the application of hybrid reinforcements to UHPFRC nullified the main advantage of using FRP to solve the corrosion problem and showed insignificant improvement in the structural performance. Synthetically, Yoo et al. (2016) recommended the use of GFRP rebar with UHPFRC, rather than the use of hybrid reinforcements.

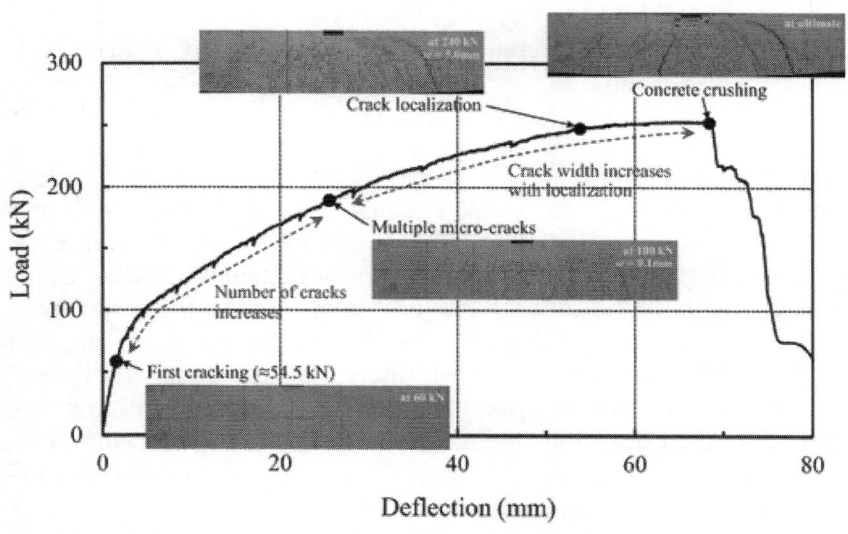

Figure 5. Typical load–deflection response of GFRP bar-reinforced UHPFRC beam (Yoo et al. 2016).

Ferrier et al. (2009) also examined the flexural behavior of a new type of hybrid beam, made of glued-laminated wood and UHPFRC planks, including steel and FRP rebar. They mention that structural efficiency was obtained by using the hybrid beams, as a consequence of the increased bending stiffness due to the high elastic modulus of UHPFRC planks. In addition, the inclusion of steel and FRP rebar in the lower UHPFRC plank significantly increased the ultimate load capacity of the hybrid beams, as compared with when only pure wood elements were used. These advantages of using hybrid beams lead to the potential for reducing the beam depth or increasing the span length of the beam, compared with conventional timber structures.

To practically apply UHPFRC in real architectural and civil structures, appropriate design technique should be suggested based on the material models. Several international recommendations (AFGC-SETRA 2002; JSCE 2004; KCI 2012) thus provide material models for designing flexural members made of UHPFRC. Based these recommendations, many researchers have already precisely predicted the flexural behaviors of reinforced UHPFRC beams (Yang et al. 2010; Yang et al. 2011; Ferrier et al.2015; Yoo and Yoon 2015; Yoo et al. 2016). In particular, the UHPFRC beams without stirrups were well predicted by AFGC-SETRA recommendations without consideration of fiber orientation coefficient ($K = 1$) (Yang et al. 2011; Yoo et al. 2016) because the fiber alignment in the direction of beam length was insignificantly disturbed by the internal rebars. However, Yoo and Yoon (2015) recently reported that the fiber orientation coefficient that is proposed by the AFGC-SETRA recommendations (i.e., $K = 1.25$), should be considered for simulating the flexural behavior of reinforced UHPFRC beams with stirrups, since the fiber orientation was clearly disturbed by the stirrups (Figure 6). In the case of FRP-reinforced concrete elements, it is well known that the service deflection prediction is the most important parameter for designing such structures, because of the larger service deflection than that of beams reinforced with steel rebar. Ferrier et al. (2015) and Yoo et al. (2016) successfully predicted the load versus deflection curves of FRP-bar-reinforced UHPFRC beams by sectional analysis, in which they considered compressive and tensile stress blocks in the cross-section, similar to the method used for the above steel-bar-reinforced beams. Yoo and Banthia (2015) also accurately predicted the service deflection of UHPFRC beams reinforced with GFRP rebar and hybrid reinforcements (steel + GFRP rebar), based on a micromechanics-based finite element (FE) analysis; the average ratios of the serviceability deflections from predictions and experiments were found to be 0.91 with a standard deviation of 0.07.

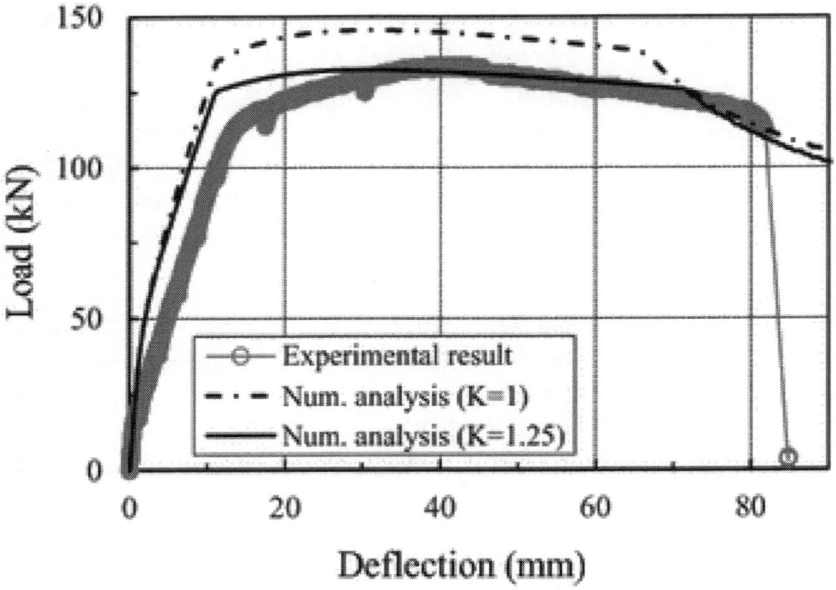

Figure 6. Comparison of experimental and numerical results for steel bar-reinforced UHPFRC beams with stirrups according to fiber orientation coefficient (Yoo and Yoon 2015).

3.3 Shear Resistance of Structural UHPFRC Beams, Girders, and Bridge Decks

Baby et al. (2013b) carried out shear tests of eleven 3-m long UHPFRC I-shaped girders with various shear reinforcements (stirrups and/or steel fibers, or neither) combined with longitudinal prestressing or passive steel bars. To examine the actual fiber orientation effect on the shear performance, the three-point flexural tests were performed by using notched prism specimens extracted from both of the undamaged ends of I-girders at different inclination angles. Test results, as shown in Figure 7, clearly indicated that the fiber orientation significantly influenced the mechanical (flexural) performance; thus, they noted that the actual fiber orientation needs to be taken into account for shear design, as recommended by AFGC-SETRA recommendations (AFGC-SETRA 2002). By including 2.5 % steel fibers, an almost 250 % increase in shear strength was observed (Baby et al. 2013c). The stirrups yielded first, while localization of the shear crack took place significantly later, as shown in Figure 8. Thus, crack localization is primarily influenced by the strain capacity of the UHPFRC, and the contributions of the fiber bridging and the stirrups up to their yield strength seem to be effective only when the tensile strain capacity of the UHPFRC is much higher than the yield strain of the stirrups. In their study (Baby et al. 2013c), the AFGC-SETRA recommendations were conservative for the shear-cracking strength, but reasonable for the ultimate shear strength prediction of UHPFRC I-girders. Baby et al. (2013a) also examined the feasibility of applying the modified compression field theory for the shear capacity of reinforced or prestressed UHPFRC beams. Based on their analytical results, the modified compression field theory was determined to be applicable for predicting the shear behavior with an effective estimation of the reorientation of the compressive struts with an increase in the load.

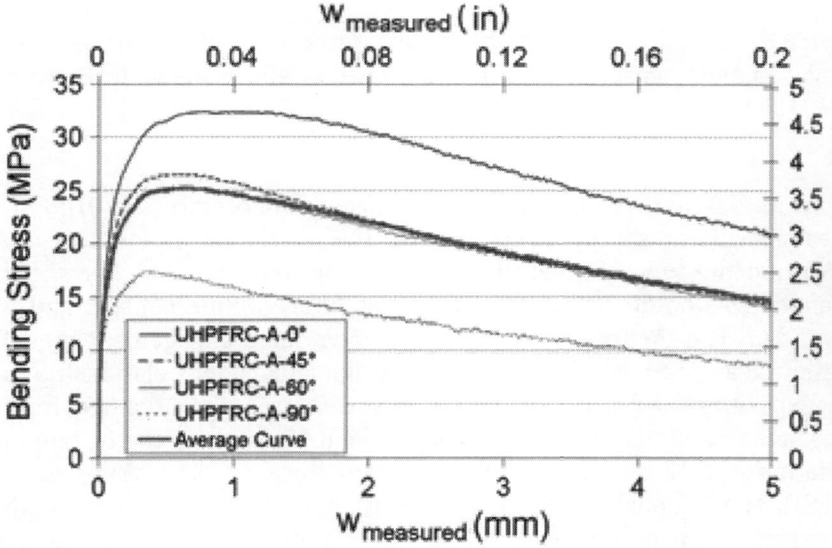

Figure 7. Three-point bending test results of prisms extracted from I-girders at different inclinations (Baby et al. 2013b).

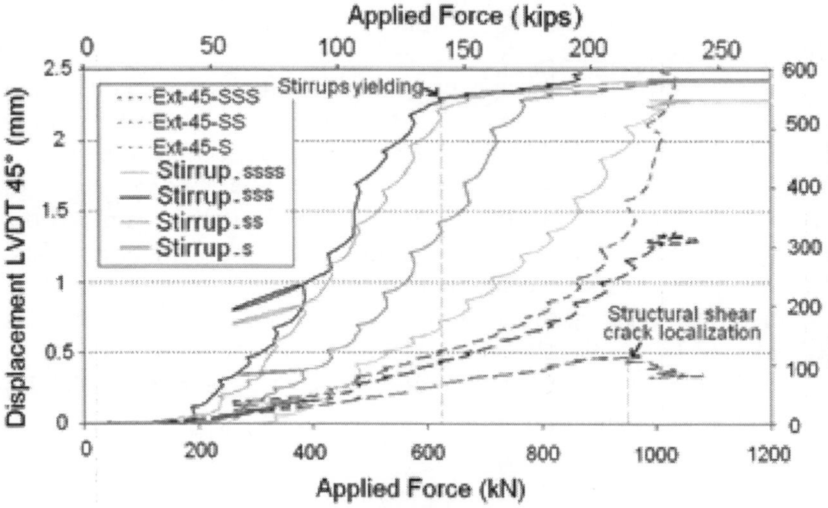

Figure 8. Evolution during tests of stresses inside stirrups and displacements measured by LVDTs attached at 45° in the web (Baby et al. 2013b).

Voo et al. (2010) investigated the shear strength of prestressed UHPFRC I-beams without stirrups, according to the shear span-to-depth ratio (a/d) and the type of steel fibers. They indicated that a higher shear strength was obtained by using a higher fiber volume content and a lower a/d. The theory of the plastic shear variable engagement model presented a good basis for their shear design and a good relationship to the experimental results; the ratio of shear strengths obtained from experiment and theory was found to be 0.92, with a coefficient of variation of 0.12. In addition, Bertram and Hegger (2012), Yang et al. (2012), and Tadepalli et al. (2015) mentioned that the shear strength increased with an increase in the fiber content and a decrease in the a/d ratio. For instance, the inclusion of 2.5 % steel fibers led to a 177 % higher ultimate load than that without fiber, and by changing the a/d ratio from 3.5 to 4.4, the shear capacity was reduced by 10 % (Bertram and Hegger 2012). Bertram and Hegger (2012) also noted that the size effect on shear strength was more substantially affected by the beam height as compared with the web thickness, and that about 12–14 % higher shear capacity was obtained when the effective prestressing force increased by 20 %. By comparing the test results with computed values, Yang et al. (2012) noticed that the predictions using the AFGC-SETRA and JSCE recommendations provided accurate estimates of the shear strength of UHPFRC I-beams (Figure 9).

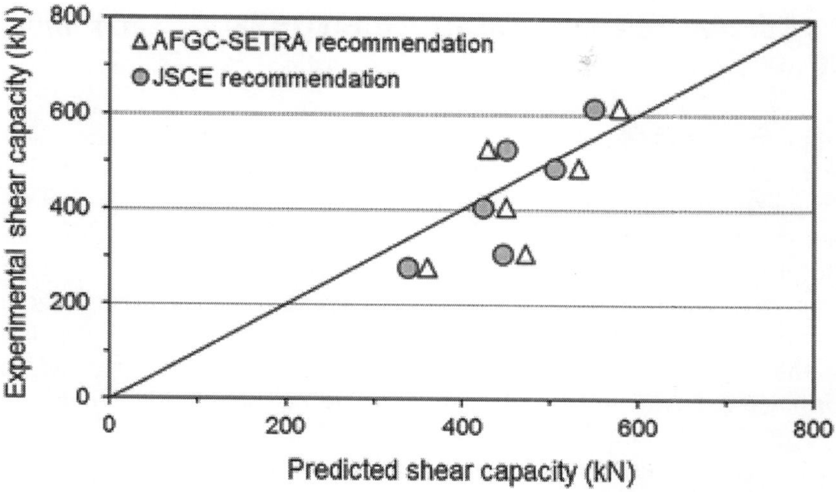

Figure 9. Comparison of experimental and predicted shear capacities (Yang et al. 2012).

In order to replace the open-grid steel decks from moveable bridges, which have several drawbacks, such as poor rideability, high noise levels, susceptibility to fatigue damage, and high maintenance costs, Saleem et al. (2011) examined the structural performance of lightweight UHPFRC bridge decks reinforced with high-strength steel rebar. They properly designed and proposed UHPFRC waffle decks to satisfy the strength, serviceability, and self-weight requirements for moveable bridges. The governing failure mode was shear, and in the multi-unit decks, shear failure was followed by punching shear failure at close to the ultimate state. However, the shear failure was less abrupt and catastrophic as compared with the commonly seen shear failure mode. Thus, Xia et al. (2011) recommended the ductile shear failure with higher post-cracking shear resistance of UHPFRC beams containing high-strength steel rebar as an acceptable failure mode, rather than including transverse reinforcements, because of their economic problems. The use of 180° hooks at both ends of the steel rebar, recommended by ACI 318 (ACI 2014), was also effective in avoiding bond failure, compared with specimens without end anchorage. Based on a thorough analysis of the experimental results, Saleem et al. (2011) noted that although the proposed UHPFRC waffle deck system exhibits shear failure mode, it has great potential to serve as an alternative to open-grid steel decks, which are conventionally used for lightweight or moveable bridges.

3.4 Torsional Behavior of Structural UHPFRC Beams and Girders

Empelmann and Oettel (2012) examined the effect of adding steel fibers ($v_f$of 1.5 and 2.5 %) on the torsional behavior of UHPFRC box girders. They experimentally observed that the inclusion of steel fibers led to a better cracking performance such as smaller crack widths and multitudinous cracks, higher ultimate and cracking torque, and improved torsional stiffness. Interestingly, the angle of the diagonal cracks was found to be approximately 45° for all test series, regardless of the steel fiber contents. Yang et al. (2013) also investigated the torsional behavior of UHPFRC beams reinforced with mild steel rebars. In order to estimate the effects of steel fiber content and transverse and longitudinal rebar ratios, thirteen UHPFRC beams were fabricated and tested. Based on their test results (Yang et al. 2013), an improvement in the initial cracking and ultimate torque were obtained by increasing the fiber volume fractions (Figure 10a), which is consistent with the findings from Empelmann and Oettel (2012). Moreover, higher ultimate torque was found with increases in the ratio of stirrups with longitudinal rebar (Figs. 10b and 11c). In addition, the torsional stiffness after initial cracking was also improved by increasing the ratio of stirrups, as shown in Figure 10b. In contrast to Empelmann and Oettel's findings, Yang et al. (2013) reported that the angle of the diagonal compressive stress ranged from 27° to 53°, and was affected by the number of stirrups and longitudinal rebar. For example, the angle of localized diagonal cracks increased with an increase in the number of stirrups, as illustrated in Figure 11. Fehling and Ismail (2012) also reported similar test results for the torsional behavior of UHPFRC elements. They specifically said that the inclusion of steel fibers was effective in improving the torsional performance, such as cracking and ultimate torsional capacities, torsional ductility, post-cracking stiffness, and toughness. The use of longitudinal reinforcements and stirrups also obviously improved the torsional performance.

3.5 Performance of Structural UHPFRC Beams, Slabs, and Columns Under Extreme Loadings

Fujikake et al. (2006a) and Yoo et al. (2015a, c) examined the impact resistance of reinforced or prestressed UHPFRC beams by testing a number of specimens using a drop-weight impact test machine. In their studies (Fujikake et al. 2006a), an increase in the maximum deflection of

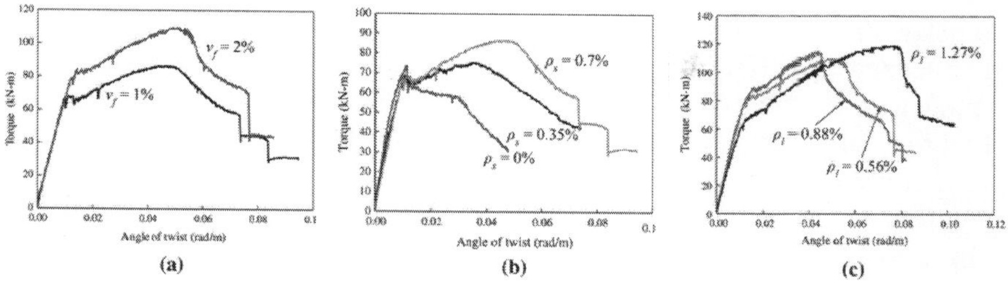

Figure 10. Torque-twist curves of UHPFRC beams (v_f = volume fraction of steel fiber, ρ_s = transverse reinforcement ratio, ρ_l = longitudinal reinforcement ratio); **a**effect of steel fiber content, **b** effect of transverse reinforcement ratio, **c** effect of longitudinal reinforcement ratio (Yang et al. 2013).

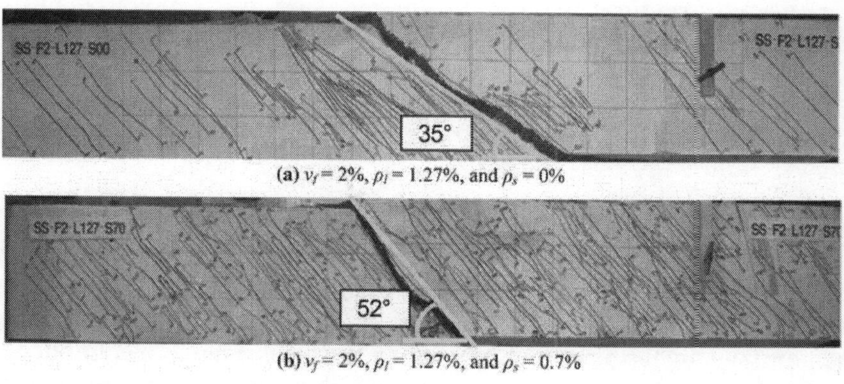

Figure 11. Angle of localized diagonal cracks (v_f = volume fraction of steel fiber, ρ_s = transverse reinforcement ratio, ρ_l = longitudinal reinforcement ratio) (Yang et al.2013).

UHPFRC beams was observed by increasing the drop height while maintaining the weight of the hammer, owing to the increase of kinetic energy. The initial stiffness in the UHPFRC beams was insignificantly affected by the impact damage because of the excellent fiber bridging capacities after matrix cracking, and the residual load–deflection (or moment–curvature) curves, shifted based on the maximum deflection by impact, exhibited quite similar behaviors with those of the virgin specimens without impact damage. Hence, Fujikake et al. (2006a) mentioned that the maximum deflection response can be used as the most rational index for estimating the overall flexural damage of reinforced UHPFRC beams. Yoo et al. (2015a) reported that better impact resistance, i.e., lower maximum and residual deflections and higher deflection recovery, was obtained by increasing the amount of longitudinal steel rebars, and the maximum and residual deflections of reinforced UHPFRC beams decreased significantly by adding 2 % (by volume) of steel fibers, leading to a change in the damage level from severe to moderate, whereas slight decreases in the maximum and residual deflections were found by increasing the fiber length at identical volume fractions (Yoo et al. 2015c). A higher ultimate load capacity was also obtained for the beams under impact loading, compared to those under quasi-static loading, and the residual load capacity after impact damage improved by including 2 % steel fibers and using the longer steel fibers. Fujikake et al. (2006a) and Yoo et al. (2015a) successfully predicted the mid-span deflection versus the time response of structural UHPFRC beams by using the sectional analysis and single- (or multi-) degree-of-freedom model. Improved mechanical compressive and tensile strengths according to the strain-rate were considered in the analysis by using the equations for the dynamic increase factor (DIF) of the UHPFRC, as suggested by Fujikake et al. (2006b, 2008).

Aoude et al. (2015) investigated the blast resistance of full-scale self-consolidating concrete (SCC) and UHPFRC columns under various blast-impulse combinations based on a shock-tube instrumentation. They verified that the steel bar-reinforced UHPFRC columns showed substantially higher blast resistance than the reinforced SCC columns in terms of reducing the maximum and residual deflections, enhancing damage tolerance, and eliminating secondary blast fragments. Based on the single-degree-of-freedom (SDOF) model and lumped inelasticity approach, Aoude et al. (2015) predicted the inelastic deflection-time histories. From the numerical results, several important findings were obtained as follows; (1) since the numerical predictions are sensitive to the choice of DIF, as given in Figure 12a, further study needs to be done to develop the strain-rate models for using in the blast analysis of UHPFRC columns, and (2) the plastic hinge length (L_p) seems to be reduced in UHPFRC columns from $L_p = d$ (column effective depth), which has been used for the analysis in conventional reinforced concrete columns, as shown in Figure 12b. Astarlioglu and Krauthammer (2014) numerically simulated the response of normal-strength concrete (NC) and UHPFRC columns subjected to blast loadings based on SDOF models using the dynamic structural analysis suite (DSAS) and reported that the UHPFRC columns presented lower mid-span displacement and sustained more than four times the impulse as compared with the NC columns.

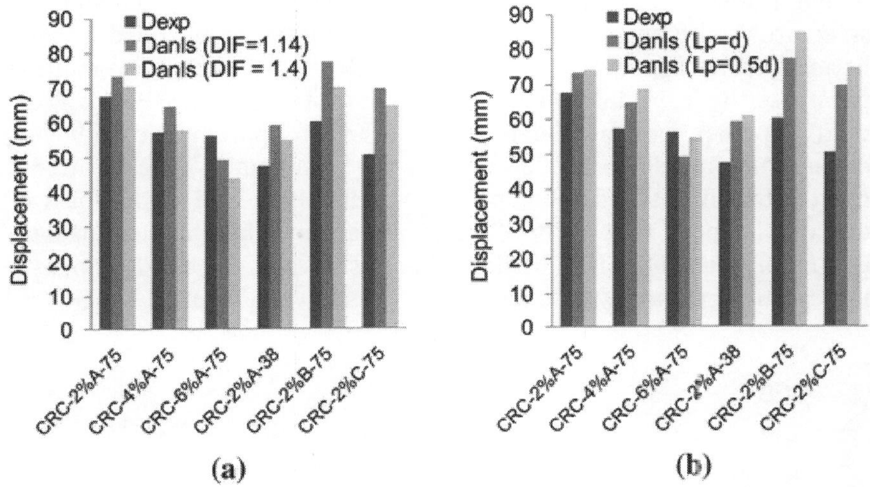

Figure 12. Displacement predictions from SDOF analysis for UHPFRC columns (*Note* CRC means UHPFRC); **a** DIF = 1.14 vs. 1.4, **b** hinge length, $L_p = d$ vs. $0.5d$(Aoude et al. 2015).

Wu et al. (2009) carried out a series of blast tests of NC and UHPFRC slabs w/ and w/o reinforcements to examine their blast resistance. When the similar blast loads were applied, the UHPFRC slabs without reinforcement exhibited less damage than the NC slabs with reinforcements, and thus, they noticed that the application of UHPFRC is effective in blast design. The UHPFRC slab with passive reinforcements was superior to all other slab specimens, and the strengthening of NC slabs with external FRP plates in the compressive zone was efficient in improving the blast resistance. Yi et al. (2012) examined the blast resistance of the reinforced slabs made of NC, ultra-high-strength concrete (UHSC), and RPC, which is identical to UHPFRC. By analyzing the crack patterns and maximum and residual deflections, they indicated that RPC has the best blast-resistant capacity, followed by UHSC and then NC. For example, the maximum deflections of NC, UHSC, and RPC slabs from 15.88 ANFO charge were found to be 18.57, 15.14, and 13.09 mm, respectively. Mao et al. (2014) investigated the capability of modeling the impact behavior of UHPFRC slabs using the commercial explicit FE program, LS-DYNA (2007). Through FE analysis, they also studied the effects of steel fibers and rebar on the blast resistance of UHPFRC slabs. Importantly, they observed that the K&C model (mostly used for simulating the blast behavior of concrete structures) with automatically generated parameters provided a much better ductile response than the actual behavior, and thus, a modified parameter b_2 from 1.35 to −2 should be applied for UHPFRC. After verifying the numerical modeling with test data, a parametric study was carried out, and some useful results were obtained: (1) the additional use of steel fibers and rebar provide similar influence in the form of extra resistance to the UHPFRC panel under far field blast loading, and (2) under near field blast loading, the resistance of the UHPFRC panels increased substantially with steel rebar, as shown in Figure 13.

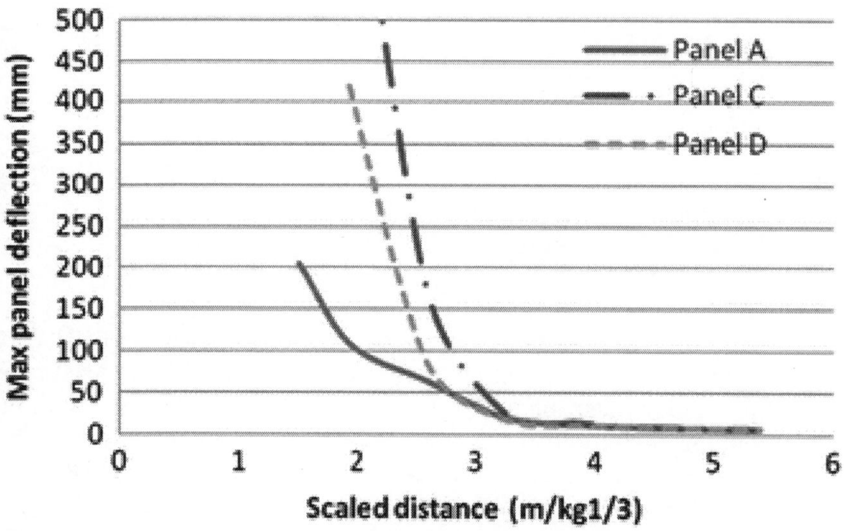

Figure 13. Variation of maximum panel deflection with scaled distance from three UHPFRC panels (*Panel A* has 2 % of 13 mm long steel fibers with 3.4 % steel rebars,*Panel C* has 2 % of 13 mm long steel fibers with 0.3 % steel rebars, and *Panel D* has 2 % of 13 mm and 2 % of 25 mm long steel fibers with 0.3 % steel rebars) (Mao et al. 2014).

4 STRUCTURAL DESIGN OF UHPFRC BASED ON AFGC-SETRA AND JSCE RECOMMENDATIONS

4.1 AFGC-SETRA Recommendations

4.1.1 Material Models

In AFGC-SETRA recommendations, UHPFRC is referred to as a cementitious material with a compressive strength in excess of 150 MPa, possibly obtaining 250 MPa, and including steel (or polymer) fibers to provide a ductile tensile behavior. The parameters of the design strength were suggested based on the mechanical test results of Ductal® (Orange et al. 1999), as follows: f_{ck} = 150–250 MPa, f_{tj} = 8 MPa, and E_c = 55 GPa, where f_{ck} is the compressive strength, f_{tj} is the post-cracking direct tensile strength, and E_c is the elastic modulus. A partial safety factor γ_{bf} is also introduced, with γ_{bf} = 1.3 in the case of fundamental combinations and γ_{bf} = 1.05 in the case of accident combinations. To consider the fiber orientation effect on the tensile behavior, three different fiber orientation coefficients were suggested, as follows: K = 1 for placement methods validated from test results of a representative model of actual structure, K = 1.25 for all loading other than local effects, and K = 1.75 for local effects.

In the case of compressive model, a bilinear stress–strain model can be used, as shown in Figure 14, with the design strength and strain parameters—$\sigma_{bcu} = 0.85 f_{ck} / \theta\gamma_b$ and $\varepsilon_u = 0.003$.

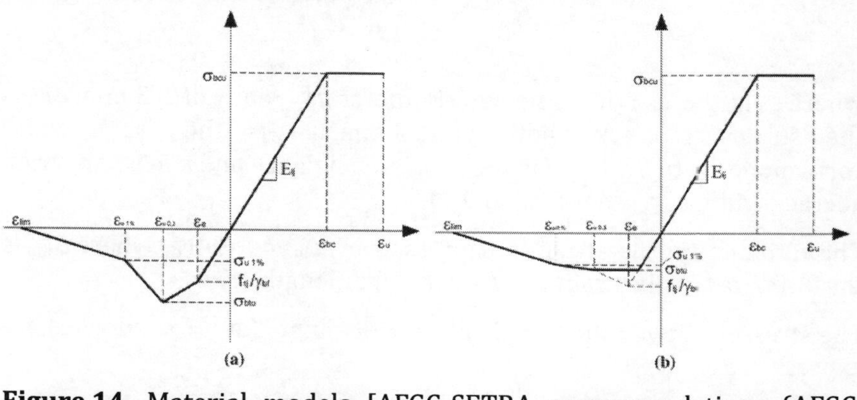

Figure 14. Material models [AFGC-SETRA recommendations (AFGC-SETRA 2002)]; **a** strain-hardening, **b** strain-softening.

The tensile stress–crack opening displacement (σ–w) model is recommended to be first derived based on an inverse analysis. To apply the σ–w model into the tensile stress block in the cross-section, it needs to be transformed to the tensile stress–strain (σ–ε) model by using the characteristic length, l_c, which is $l_c = 2/3 \times h$ for the case of rectangular or T-beams, where h is the beam height.

To obtain the tensile σ–ε model, the elastic tensile strain and strains at crack widths of 0.3 mm and 1 % of beam height need to be calculated based on the following equations:

$$\varepsilon_e = \frac{f_{tj}}{E_c} \tag{5}$$

$$\varepsilon_{0.3} = \frac{w_{0.3}}{l_c} + \frac{f_{tj}}{\gamma_{bf} E_c} \tag{6}$$

$$\varepsilon_{1\%} = \frac{w_{1\%}}{l_c} + \frac{f_{tj}}{\gamma_{bf} E_c} \tag{7}$$

where ε_e is the elastic strain, $w_{0.3}$ is the crack width of 0.3 mm, $\varepsilon_{0.3}$ is the strain at crack width of 0.3 mm, $w_{1\%}$ is the crack width corresponding to 0.01H (H is the specimen height), and $\varepsilon_{1\%}$ is the strain at crack width corresponding to 0.01H.

The ultimate tensile strain is expressed by $\varepsilon_{\lim} = L_f/4l_c$, where $\varepsilon_{\lim}$ is the ultimate tensile strain and L_f is the fiber length.

The stresses at two different characteristic points (at $w_{0.3}$ and $w_{1\%}$) are expressed as follows:

$$f_{bt} = \frac{f(w_{0.3})}{K\gamma_{bf}} \tag{8}$$

$$f_{1\%} = \frac{f(w_{1\%})}{K\gamma_{bf}} \tag{9}$$

where f_{bt} is the stress at a crack width of 0.3 mm and $f_{1\%}$ is the stress at a crack width corresponding to $0.01H$.

The completed material models under compression and tension are given in Figure 14. Based on the capacity of tensile resistance with crack opening displacement, the AFGC-SETRA recommendations classify the tensile response by two different laws: (1) the strain-softening law ($f_{tj} > f_{bt}$) and (2) the strain-hardening law ($f_{tj} < f_{bt}$), as shown in Figure 14.

Other important material properties of UHPFRC with heat treatment are given by AFGC-SETRA recommendations, as follows:

- Poisson's ratio: 0.2.
- Thermal expansion coefficient ($\times 10^{-6}$/°C): 11.0.
- Long-term creep coefficient: 0.2.
- Total (autogenous) shrinkage: 550×10^{-6}.

4.1.2 Flexural Design

A chapter for the flexural design of UHPFRC structures is not included in the AFGC-SETRA recommendations. However, based on the suggested material models in Figure 14, the design of a UHPFRC element subjected to bending can be performed by using the sectional analysis (Fujikake et al. 2006a; Yang et al. 2011; Ferrier et al. 2015; Yoo and Yoon 2015; Yoo et al. 2016). A schematic description of a multi-layered cross-section with strain and stress distributions and an algorithm for sectional analysis are shown in Figure 15 (Yoo and Yoon 2015). The cross-section

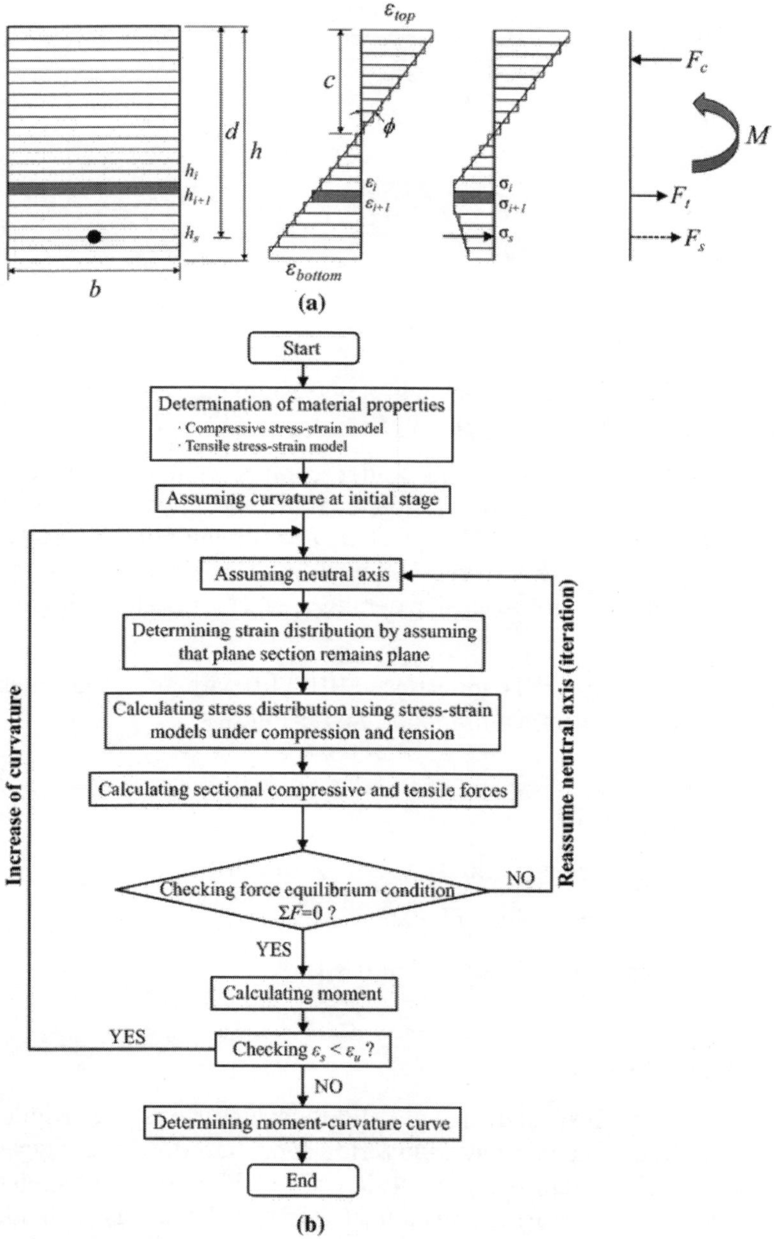

Figure 15. Sectional analysis; **a** schematic description of stress and strain distributions in cross-section, **b** algorithm for sectional analysis (Yoo and Yoon 2015).

of the elements is first divided into a number of layers along the height. After that, the compressive and tensile stresses at each layer can be calculated by assuming that the plane section remains a plane at a given curvature. Then, the neutral axis depth can be calculated from the force equilibrium condition. Lastly, the moment is calculated. The calculation was repeated until the ultimate strain of the steel rebar was reached.

4.1.3 Shear Design

Since the steel fibers can resist a portion of the stress at shear cracks, they mentioned that stirrups may be used, but the shear strength given by the fibers may make it possible to dispense with the stirrups. The ultimate shear strength V_u is given by:

$$V_u = V_{Rb} + V_a + V_f \quad (10)$$

where V_{Rb} is the term for the participation of the concrete, V_a is the term for the participation of the reinforcement, and V_f is the term for the participation of the fibers.

The shear strength by reinforcement V_a is given by:

$$V_a = 0.9d \frac{A_t f_e}{s_t \gamma_s} (\sin \alpha + \cos \alpha) \quad (11)$$

where d is the effective depth, A_t is the area of the stirrups, s_t is the spacing of the stirrups, f_e is the stress in the stirrups, γ_s is the partial safety factor, and α is the inclination angle of the stirrups.

V_{Rb} is expressed by two equations for reinforced and prestressed concrete, as follow:

$$V_{Rb} = \frac{1}{\gamma_E} \frac{0.21}{\gamma_b} k \sqrt{f_{ck}} b_0 d \quad \text{(for reinforced concrete)} \quad (12)$$

$$V_{Rb} = \frac{1}{\gamma_E} \frac{0.24}{\gamma_b} \sqrt{f_{ck}} b_0 z \quad \text{(for prestressed concrete)} \quad (13)$$

where γ_E is the safety coefficient such that: $\gamma_E \times \gamma_b = 1.5$, b_0 is the element width, σ_m is the mean stress in the total section of concrete under the normal design force, $k = 1 + 3\sigma_{cm}/f_{tj}$ for compression, $k = 1 - 0.7\sigma_{tm}/f_{tj}$ for tension, and z is the distance from the top fiber to the center of prestressing strand.

Lastly, the shear strength of the fibers is calculated by using the following equation:

$$V_f = \frac{S\sigma_p}{\gamma_{bf} \tan \beta_u} \tag{14}$$

$$\sigma_p = \frac{1}{K}\frac{1}{w_{\lim}} \int_0^{w_{\lim}} \sigma(w)dw \tag{15}$$

with $w_{lim} = \max(w_u;\ 0.3\text{ mm})$ and $w_u = l_c\varepsilon_u$

where S is the area of the fiber effect ($S = 0.9b_0 d$ or $b_0 d$ for rectangular or T-sections, and $S = 0.8(0.9d)^2$ or $0.8z^2$ for circular sections), β_u is the inclination angle between a diagonal crack and the longitudinal direction of the beam, K is the orientation coefficient for general effects, $\sigma(w)$ is the experimental characteristic post-cracking stress for a crack width of w, and w_u is the ultimate crack width.

4.2 JSCE Recommendations

4.2.1 Material Models

In JSCE recommendations, UHPFRC is defined as the concrete with $f_c' \geq 150$ MPa, $f_{crk} \geq 4$ MPa, and $f_{tk} \geq 5$ MPa, where f_c' is the compressive strength, f_{crk} is the cracking strength, and f_{tk} is the tensile stress at crack width of 0.5 mm. JSCE recommendations were proposed based on Ductal® (Orange et al. 1999), which is commercially available UHPFRC with heat treatment and 2 vol.% of steel fibers having $d_f = 0.2$ mm and $L_f = 15$ mm, and provided the strength properties used for structural design as follows; $f_c' = 180$ MPa, $f_{crk} = 8$ MPa, $f_{tk} = 8.8$ MPa, and $E_c = 50$ GPa.

Importantly, they suggested a bilinear stress–strain curve for compression (Figure 16a) and a bilinear tension-softening curve (TSC) with f_{tk} = 8.8 MPa, w_{1k} = 0.5 mm, and w_{2k} = 4.3 mm for tension (Figure 16b). In order to obtain the structural safety, γ_c = 1.3 was proposed for the partial safety factor. To take into account the suggested TSC for tensile stress block in the cross-section, the crack opening displacement should be transformed to a strain by using the equivalent specific length, L_{eq}, as follows

$$L_{eq} = 0.8h\left[1 - \frac{1}{\left(1.05 + {}^{6h}/_{l_{ch}}\right)^4}\right] \quad (16)$$

where h is the overall depth of beam, l_{ch} is the characteristic length $(=G_F E_c / f_{tk}^2)$, and G_F is the fracture energy.

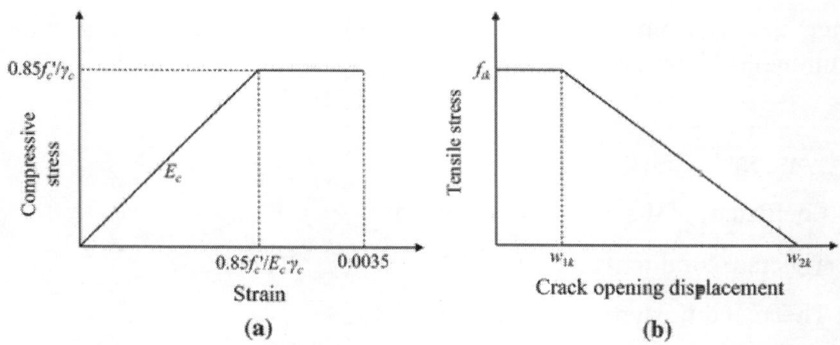

Figure 16. Material models [JSCE recommendations (JSCE 2004)]; **a** compressive stress–strain curve, **b** tension-softening curve.

By using the equivalent specific length, the tensile stress–strain model is obtained by the following equations, based on the TSC:

$$\varepsilon_{cr} = \frac{f_{tk}}{\gamma_c E_c} \tag{17}$$

$$\varepsilon_1 = \varepsilon_{cr} + \frac{w_{1k}}{L_{eq}} \tag{18}$$

$$\varepsilon_2 = \frac{w_{2k}}{L_{eq}} \tag{19}$$

where ε_{cr} is the factored elastic strain, ε_1 is the strain at the end of the initial plateau, w_{1k} is the crack opening displacement for which a certain stress level is retained after the first crack, ε_2 is the strain at zero tensile stress, and w_{2k} is the crack opening displacement at zero tensile stress.

Other material properties of the UHPFRC, proposed by the JSCE recommendations, are as follows:

- Poisson's ratio: 0.2.
- Coefficient of thermal expansion ($\times 10^{-6}$/°C): 13.5.
- Thermal conductivity (kJ/mh °C): 8.3.
- Thermal diffusivity ($\times 10^{-3} m^2/h$): 3.53.
- Specific heat (kJ/kg °C): 0.92.
- Total shrinkage: 550×10^{-6}.
- Creep coefficient: 0.4.
- Density (kN/m^3) for calculating dead load: 25.5.

4.2.2 Flexural Design

In the JSCE recommendations, for the structural design of UHPFRC elements under bending, two simple assumptions are required to be satisfied: (1) the linear strain distribution and (2) the use of the proposed material models, as given in Figure 16. The steel fiber contribution in the tensile zone after cracking needs to be considered in the structural design, and the compressive and tensile stress blocks in the cross-section should be considered based on the proposed material stress–strain models (Figure 16) by considering the equivalent specific length for tension. Although a detailed procedure for calculating the ultimate moment capacity is not introduced in the JSCE recommendations, sectional analysis can be adopted for calculating the moment–curvature behavior of the UHPFRC elements (Yang et al. 2011; Yoo et al. 2016), similar to the case of the AFGC-SETRA recommendations.

4.2.3 Shear Design

For the shear design of the UHPFRC elements, the shear resistance from the matrix and the included steel fibers is required to be calculated. In accordance with the JSCE recommendations, the total shear resistance can be calculated by:

$$V_{yd} = V_{rpcd} + V_{fd} + V_{ped} \tag{20}$$

where V_{yd} is the total shear resistance of the reinforced UHPFRC beams, V_{rpcd} is the shear resistance of the matrix without fiber, V_{fd} is the shear resistance of the steel fibers, and V_{ped} is the shear resistance of the stirrups.

Herein, V_{rpcd} and V_{ped} are obtained by using Eqs. (21) and (22), respectively.

$$V_{rpcd} = 0.18\sqrt{f'_c} b_w d / \gamma_b \tag{21}$$

$$V_{fd} = (f_{vd} / \tan \beta_u) b_w z / \gamma_b \tag{22}$$

where b_w is the web width of the beam, d is the effective depth of the beam, γ_b is the strength reduction factor (γ_b = 1.3), f_{vd} is the design tensile strength perpendicular to the diagonal crack (= f_{tk} / γ_b), z is the distance between the point-acting compressive force and the center tensile reinforcement (=d/1.15), and β_u is the inclination angle between the diagonal crack and the longitudinal direction of the beam. The inclination angle β_u should be larger than 30° and is calculated by $\beta_u = 1/2\tan^{-1}[2\tau/(\sigma_{xu} - \sigma_{yu})] - \beta_0$, where τ is the average shear stress, σ_{xu} and σ_{yu} are the average compressive stress in longitudinal and transverse directions, and β_0 is the inclination angle without axial force.

In addition, V_{ped} can be calculated as follows

$$V_{ped} = P_{ed} \sin \alpha_p / \gamma_b \tag{23}$$

where P_{ed} is the effective tensile force of the prestressing strand, α_p is the inclination angle between the prestressing strand and longitudinal axis of beam, and γ_b is the strength reduction factor (γ_b = 1.1).

As was reported by Yang et al. (2012), the shear design of UHPFRC elements can be carried out by using both AFGC-SETRA and JSCE recommendations, which provide good estimates with test data of I-shaped UHPFRC beams, as shown in Figure 9.

5 FIELD APPLICATIONS OF UHPFRC

Due to its excellent mechanical performance, UHPFRC can lead to a reduction in the number of sections, the elimination of passive reinforcements, and the possibility for the design of structures that is not possible with ordinary concrete (NPCA 2011). For this reason, UHPFRC has attracted much attention from engineers for field applications from 1995 to 2010 (Voo et al. 2012). The representative application examples in North America, Europe, and Asia are briefly explained herein. The first structural application of UHPFRC was the prestressed hybrid pedestrian bridge at Sherbrooke in Canada, constructed in 1997 (Resplendino 2004) (Figure 17a). This precast and prestressed pedestrian bridge includes a post-tensioned open-web space UHPFRC truss with 4 access spans made by conventional high-performance concrete. A ribbed slab with a 30-mm thickness was adopted, and a transverse prestressing was applied with sheathed monostrands. The UHPFRC truss webs were confined by steel tubes, and the structure was longitudinally prestressed by both internal and external prestressing strands. The total span length of the bridge was 60 m, and the main span was assembled from six 10-m prefabricated match-cast segments. The Bourg-lès-Valence bridge was the first UHPC road bridge in the world, built in France in 2001 (Hajar et al. 2004), as shown in Figure 17b. The bridge was built from five assembled π-shaped precast UHPFRC beams, and the joints were made by in situ UHPFRC with internal reinforcements. The bridge consisted of two isolated spans with a length of approximately 20 m, and all the π-shaped beams were prestressed without transverse passive reinforcement. The Seonyu Footbridge, completed in 2002 in Seoul, Korea, is currently the longest footbridge made by UHPFRC (Ductal®) with a single span of 120 m and no central support (Figure 17c). It consists of a π-shaped arch supporting a ribbed UHPFRC slab with a thickness of 30 mm, and transverse prestressing was provided by sheathed monostrands. With equivalent load carrying capacity and strength properties, the bridge needed only half of the amount of materials required for conventional concrete construction (Voo et al. 2014). The first UHPFRC highway bridge, built in Iowa, USA, is the Mars Hill bridge, as shown in Figure 17d. This is a simple single-span bridge consisting of three precast and prestressed concrete beams with a length of 33.5 m (modified 1.14-m-deep Iowa bulb-tee beams), and the cast-in-place concrete bridge deck was topped. No stirrup was applied, and each beam included 47 low-relaxation prestressing strands with a diameter of 15.2 mm (Russell and Graybeal 2013).

Curved UHPFRC panels were applied to the building named the Atrium, which was built in Victoria, BC, Canada in 2013 (Figure 18a) (NPCA 2011). In this project, UHPFRC was selected as an appropriate material for the curved panel system, due to its ability to form monolithically tight radial curves, and consequently, it could improve the energy efficiency of the building by eliminating unattractive seams and openings. The Community Center in Sedan, France was built in 2008 with a double skin facade to protect the glass fascia behind and to provide privacy using light blue UHPFRC perforated panels (Figure 18b). They designed the UHPFRC panels to be 2 m by 4 m by 45 mm thick, to permit sunlight to stream through to the interior spaces. The main reason for choosing UHPFRC panels instead of traditional perforated panels, made of metal, painted steel, cast aluminum, cast iron, and stainless steel, was that UHPFRC is durable, and requires less energy consumption for fabrication and for maintenance over time (Henry et al. 2011).

Figure 17. Examples of UHPFRC applications in bridges.

Figure 18. Examples of UHPFRC applications in buildings.

Mega-architectural projects were carried out to build the Stade Jean Bouin and the MuCEM in France (Figure 19) (NPCA 2011). In the case of the Stade Jean Bouin, the first application of a precast UHPFRC lattice-style facade system in the world was made. The 23,000 m^2 envelope, which contains a 12,000 m^2 roof, was built from 3600 self-supporting UHPFRC panels that are 8–9 m long by 2.5 m wide, and 45 mm thick. The MuCEM consists of several types of precast UHPFRC structural elements, such as 78-m-long footbridge, main cubic structures with a

15,000 m² surface area, flooring, and lattice-style envelope with a series of slender N- and Y-shaped columns. The architect designed the MuCEM to carry the entire external load by columns, and to satisfy the maximum requirements regarding both seismic and fire resistance, as specified by CSTB.

Figure 19. Mega architectural projects in France (NPCA 2011).

6 CONCLUSION

This paper reviewed the current state-of-the-art for structural performance, design recommendations, and applications of UHPFRC. From the above literature review and discussions, the following conclusions are drawn:

1. The use of 2 % steel fibers resulted in a higher post-cracking stiffness and ultimate load capacity of steel bar-reinforced UHPFRC beams, but it decreased their ductility because of the superb bond strength with steel bars and crack localization characteristics. Importantly, the use of UHPFRC was effective in overcoming the major drawbacks of conventional FRP-reinforced concrete structures (large service deflection) due to its strain-hardening response, and the use of CFRP bars was efficient in improving the flexural stiffness of reinforced UHPFRC beams, compared to that of GFRP bars. The hybrid reinforcements (steel + FRP bars), which have been adopted to reduce the service deflection of conventional FRP-reinforced

beams, were ineffective in UHPFRC beams, and thus, the use of single FRP bars, instead of hybrid reinforcements, was recommended for the case of UHPFRC.

2. With the inclusion of 2.5 % steel fibers, approximately 250 % higher shear strength was obtained, compared to that without fibers. The shear strength also increased with an increase in the fiber contents and a decrease in the shear span-to-depth ratio. Due to the excellent fiber bridging at crack surfaces, the shear failure of UHPFRC beams was less abrupt than the commonly seen shear failure mode in conventional concrete. In addition, both AFGC-SETRA and JSCE recommendations provided accurate predictions of shear strength of UHPFRC beams.

3. The inclusion of steel fibers provided a better cracking performance, higher ultimate and cracking torque, and improved torsional stiffness. The higher ultimate torque was also found by increasing the ratios of stirrups and longitudinal rebars. The diagonal crack angle was influenced by the amount of stirrups and longitudinal rebars, whereas it was not affected by the amount of steel fibers.

4. Better impact resistance of reinforced UHPFRC beams was obtained by including 2 % steel fibers and a larger number of longitudinal steel bars. The residual capacity after impact damage was also improved by adding 2 % steel fibers and using the longer steel fibers. UHPFRC columns exhibited significantly higher blast resistance such as lower maximum deflection, improved damage tolerance, and higher resistance, compared with the reinforced SCC and NC columns. Thus, it was noted that the application of UHPFRC in the structures subjected to blast loads is effective.

5. Finally, the international design recommendations on UHPFRC were discussed minutely, and examples of practical applications of UHPFRC in architectural and civil structures were investigated.

ACKNOWLEDGMENTS

This research was supported by a grant from a Construction Technology Research Project 13SCIPS02 (Development of impact/blast resistant HPFRCC and evaluation technique thereof) funded by the Ministry of Land, Infrastructure and Transport.

REFERENCES

1. ACI Committee 440. (2006). *Guide for the design and construction of concrete reinforced with FRP bars (ACI 440.1R-06)*. Farmington Hills, MI: American Concrete Institute.
2. AFGC-SETRA. (2002). *Ultra high performance fibre-reinforced concretes. Interim recommendations*. Bagneux, France: SETRA.
3. Ahmad Firas, S., Foret, G., & Le Roy, R. (2011). Bond between carbon fibre-reinforced polymer (CFRP) bars and ultra high performance fibre reinforced concrete (UHPFRC): Experimental study. *Construction and Building Materials, 25*(2), 479–485.
4. American Concrete Institute (ACI). (2014). *Building code requirements for structural concrete and commentary. ACI 318-14 and ACI 318R-14*. Farmington Hills, MI: American Concrete Institute (ACI).
5. Aoude, H., Dagenais, F. P., Burrell, R. P., & Saatcioglu, M. (2015). Behavior of ultra-high performance fiber reinforced concrete columns under blast loading.*International Journal of Impact Engineering, 80*, 185–202.
6. Astarlioglu, S., & Krauthammer, T. (2014). Response of normal-strength and ultra-high-performance fiber-reinforced concrete columns to idealized blast loads.*Engineering Structures, 61*, 1–12.
7. Baby, F., Marchand, P., Atrach, M., & Toutlemonde, F. (2013a). Analysis of flexure-shear behavior of UHPFRC beams based on stress field approach. *Engineering Structures, 56*, 194–206.
8. Baby, F., Marchand, P., & Toutlemonde, F. (2013b). Shear behavior of ultrahigh performance fiber-reinforced concrete

beams. I: Experimental investigation. *Journal of Structural Engineering, 140*(5), 04013111.

9. Baby, F., Marchand, P., & Toutlemonde, F. (2013c). Shear behavior of ultrahigh performance fiber-reinforced concrete beams. II: Analysis and design provisions. *Journal of Structural Engineering, 140*(5), 04013112.

10. Bache, H. H. (1981). Densified cement ultra-fine particle-based materials. In *Proceedings of the 2nd international conference on superplasticizers in concrete*, Ottawa, Canada, p. 33.

11. Bertram, G., & Hegger, J. (2012). Shear behavior of pretensioned UHPC beams—Tests and design. In *Proceedings of the third international symposium on UHPC and nanotechnology for high performance construction materials*, Kassel, pp. 493–500.

12. Birchall, J. D., Howard, A. J., & Kendall, K. (1981). Flexural strength and porosity of cements. *Nature, 289*(5796), 388–390.

13. CAN/CSA S806. (2002). *Design and construction of building components with fibre reinforced polymers*, Rexdale, Canada.

14. CAN/CSA-S6. (2006). *Canadian highway bridge design code*, Toronto, ON.

15. CEB-FIP. (1993). *Model code for concrete structures*. CEB Bulletin d'Information. Comite Euro international du Beton, Lausanne, Switzerland.

16. Choi, W. C., Yun, H. D., Cho, C. G., & Feo, L. (2014). Attempts to apply high performance fiber-reinforced cement composite (HPFRCC) to infrastructures in South Korea. *Composite Structures, 109*, 211–223.

17. Cosenza, E., Manfredi, G., & Realfonzo, R. (1995). Analytical modelling of bond between FRP reinforcing bars and concrete. In L. Taerwe (Ed.), *Proceedings of second international RILEM symposium (FRPRCS-2)* (pp. 164–171). London, UK: E and FN Spon.

18. DAfStB UHPC. (2003). *State-of-the-art report on ultra high performance concrete—Concrete technology and design*. Deutscher Ausschuss für Stahlbeton/German Association for Reinforced Concrete, Berlin, Germany, draft 3.

19. Dancygier, A. N., & Berkover, E. (2016). Cracking localization and reduced ductility in fiber-reinforced concrete beams with low reinforcement ratios. *Engineering Structures, 111*, 411–424.

20. Empelmann, M., & Oettel, V. (2012). UHPFRC box girders under torsion. In *Proceedings of the third international symposium on UHPC and nanotechnology for high performance construction materials*, Kassel, Germany, pp. 517–524.

21. Farhat, F. A., Nicolaides, D., Kanellopoulos, A., & Karihaloo, B. L. (2007). High performance fibre-reinforced cementitious composite (CARDIFRC)—Performance and application to retrofitting. *Engineering Fracture Mechanics, 74*(1–2), 151–167.

22. Fehling, E., & Ismail, M. (2012). Experimental investigations on UHPC structural elements subjected to pure torsion. In *Proceedings of the third international symposium on UHPC and nanotechnology for high performance construction materials*, Kassel, Germany, pp. 501–508.

23. Ferrier, E., Labossiere, P., & Neale, K. W. (2009). Mechanical behavior of an innovative hybrid beam made of glulam and ultrahigh-performance concrete reinforced with FRP or steel. *Journal of Composites for Construction, 14*(2), 217–223.

24. Ferrier, E., Michel, L., Zuber, B., & Chanvillard, G. (2015). Mechanical behaviour of ultra-high-performance short-fibre-reinforced concrete beams with internal fibre reinforced polymer bars. *Composites Part B: Engineering, 68*, 246–258.

25. Fujikake, K., Senga, T., Ueda, N., Ohno, T., & Katagiri, M. (2006a). Study on impact response of reactive powder concrete beam and its analytical model. *Journal of Advanced Concrete Technology, 4*(1), 99–108.

26. Fujikake, K., Senga, T., Ueda, N., Ohno, T., & Katagiri, M. (2006b). Effects of strain rate on tensile behavior of reactive powder concrete. *Journal of Advanced Concrete Technology, 4*(1), 79–84.

27. Fujikake, K., Uebayashi, K., Ohno, T., Shimoyama, Y., & Katagiri, M. (2008). Dynamic properties of steel fiber reinforced mortar under high-rates of loadings and triaxial stress states. In *Proceedings of the 7th international conference on structures under shock and impact* (pp. 437–446). Montreal, Canada: WIT Press.

28. Graybeal, B. A. (2008). Flexural behavior of an ultrahigh-performance concrete I-girder. *Journal of Bridge Engineering, 13*(6), 602–610.

29. Graybeal, B., & Tanesi, J. (2007). Durability of an ultrahigh-performance concrete. *Journal of Materials in Civil Engineering, 19*(10), 848–854.

30. Hajar, Z., Lecointre, D., Simon, A., & Petitjean, J. (2004). Design and construction of the world first ultra-high performance concrete road bridges. In *Proceeding of the international symposium on ultra high performance concrete*, University of Kassel, Kassel, Germany, pp. 39–48.

31. Henry, K. A., Seibert, P. J., & America, L. N. (2011). Manufacturing UHPC architectural products. http://www.ductal.fr/CPI_-_OCTOBER_2011.pdf.

32. Jaeger, G. L., Tadros, G., & Mufti, A. A. (1995). *Balanced section, ductility and deformability in concrete with FRP reinforcement.* Technical report no. 2-1995, Nova Scotia computer aided design/computer aided manufacturing centre, Technical University of Nova Scotia, Halifax, Canada.

33. JSCE. (2004). *Recommendations for design and construction of ultra-high strength fiber reinforced concrete structures (Draft).* Tokyo, Japan: Japan Society of Civil Engineers.

34. Jungwirth, J., & Muttoni, A. (2004). Structural behavior of tension members in UHPC. In *Proceedings of the international symposium on ultra-high-performance concrete,* Kassel, Germany, pp. 533–544.

35. KCI. (2012). *Design recommendations for ultra-high performance concrete K-UHPC. KCI-M-12-003.* Seoul: Korea Concrete Institute.

36. Kim, S. W., Park, J. J., Kang, S. T., Ryo, G. S., & Koh, K. T. (2008). Development of ultra high performance cementitious composites (UHPCC) in Korea. In*Proceedings of the fourth international IABMAS conference,* Seoul, Korea, p. 110.

37. Lau, D., & Pam, H. J. (2010). Experimental study of hybrid FRP reinforced concrete beams. *Engineering Structures, 32*(12), 3857–3865.

38. Li, H., & Liu, G. (2016). Tensile properties of hybrid fiber-reinforced reactive powder concrete after exposure to elevated

temperatures. *International Journal of Concrete Structures and Materials, 10*(1), 29–37.

39. Livermore Software Technology Corporation. (2007). *LS-DYNA user's manual—Version 971*. Livermore, CA: Livermore Software Technology Corporation.

40. Mao, L., Barnett, S., Begg, D., Schleyer, G., & Wight, G. (2014). Numerical simulation of ultra high performance fibre reinforced concrete panel subjected to blast loading. *International Journal of Impact Engineering, 64*, 91–100.

41. NPCA White Paper. (2011). Ultra high performance concrete (UHPC), guide to manufacturing architectural precast UHPC elements. http://precast.org/ wp-content/uploads/2011/05/NPCA-ultra-high-performance-concrete.pdf

42. Orange, G., Acker, P., & Vernet, C. (1999). A new generation of UHP concrete: Ductal® damage resistance and micromechanical analysis. In *Proceedings of the third international workshop on high performance fiber reinforced cement composites (HPFRCC3)*, Mainz, Germany, pp. 101–111.

43. Resplendino, J. (2004). First recommendations for ultra-high-performance concretes and examples of application. In *Proceeding of the international symposium on ultra high performance concrete*, University of Kassel, Kassel, Germany, pp. 79–90.

44. Richard, P., & Cheyrezy, M. (1995). Composition of reactive powder concretes. *Cement and Concrete Research, 25*(7), 1501–1511.

45. RILEM TC. (1994). *RILEM recommendations for the testing and use of constructions materials. RC 6 bond test for reinforcement steel. 2. Pull-out test, 1983* (pp. 218–220). London, UK: E & FN SPON.

46. Roy, D. M., Gouda, G. R., & Bobrowsky, A. (1972). Very high strength cement pastes prepared by hot pressing and other high pressure techniques. *Cement and Concrete Research, 2*(3), 349–366.

47. Russell, H. G., & Graybeal, B. A. (2013). *Ultra-high performance concrete: A state-of-the-art report for the bridge community, FHWA-HRT-13-060*.

48. Saleem, M. A., Mirmiran, A., Xia, J., & Mackie, K. (2011). Ultra-high-performance concrete bridge deck reinforced with high-strength steel. *ACI Structural Journal,108*(5), 601–609.

49. Schäfers, M., & Seim, W. (2011). Investigation on bonding between timber and ultra-high performance concrete (UHPC). *Construction and Building Materials,25*(7), 3078–3088.

50. Tadepalli, P. R., Dhonde, H. B., Mo, Y. L., & Hsu, T. T. (2015). Shear strength of prestressed steel fiber concrete I-beams. *International Journal of Concrete Structures and Materials, 9*(3), 267–281.

51. Tam, C. M., Tam, V. W., & Ng, K. M. (2012). Assessing drying shrinkage and water permeability of reactive powder concrete produced in Hong Kong. *Construction and Building Materials, 26*(1), 79–89.

52. Voo, Y. L., Foster, S. J., & Voo, C. C. (2014). Ultrahigh-performance concrete segmental bridge technology: Toward sustainable bridge construction. *Journal of Bridge Engineering, 20*(8), B5014001.

53. Voo, Y. L., Nematollahi, B., Said, A., Gopal, A., & Yee, T. Y. (2012). Application of ultra high performance fiber reinforced concrete—The Malaysia perspective. *International Journal of Sustainable Construction Engineering and Technology, 3*(1), 26–44.

54. Voo, Y. L., Poon, W. K., & Foster, S. J. (2010). Shear strength of steel fiber-reinforced ultrahigh-performance concrete beams without stirrups. *Journal of Structural Engineering, 136*(11), 1393–1400.

55. Wambeke, B. W., & Shield, C. K. (2006). Development length of glass fiber-reinforced polymer bars in concrete. *ACI Structural Journal, 103*(1), 11–17.

56. Wu, C., Oehlers, D. J., Rebentrost, M., Leach, J., & Whittaker, A. S. (2009). Blast testing of ultra-high performance fibre and FRP-retrofitted concrete slabs. *Engineering Structures, 31*(9), 2060–2069.

57. Xia, J., Mackie, K. R., Saleem, M. A., & Mirmiran, A. (2011). Shear failure analysis on ultra-high performance concrete beams reinforced with high strength steel. *Engineering Structures, 33*(12), 3597–3609.

58. Yang, I. H., Joh, C., & Kim, B. S. (2010). Structural behavior of ultra high performance concrete beams subjected to bending. *Engineering Structures, 32*(11), 3478–3487.

59. Yang, I. H., Joh, C., & Kim, B. S. (2011). Flexural strength of large-scale ultra high performance concrete prestressed T-beams. *Canadian Journal of Civil Engineering, 38*(11), 1185–1195.

60. Yang, I. H., Joh, C., Lee, J. W., & Kim, B. S. (2012). An experimental study on shear behavior of steel fiber-reinforced ultra high performance concrete beams.*Journal of The Korean Society of Civil Engineers, 32*(1A), 55–64.

61. Yang, I. H., Joh, C., Lee, J. W., & Kim, B. S. (2013). Torsional behavior of ultra-high performance concrete squared beams. *Engineering Structures, 56*, 372–383.

62. Yi, N. H., Kim, J. H. J., Han, T. S., Cho, Y. G., & Lee, J. H. (2012). Blast-resistant characteristics of ultra-high strength concrete and reactive powder concrete.*Construction and Building Materials, 28*(1), 694–707.

63. Yoo, D. Y., & Banthia, N. (2015). Numerical simulation on structural behavior of UHPFRC beams with steel and GFRP bars. *Computers and Concrete, 16*(5), 759–774.

64. Yoo, D. Y., Banthia, N., Kim, S. W., & Yoon, Y. S. (2015a). Response of ultra-high-performance fiber-reinforced concrete beams with continuous steel reinforcement subjected to low-velocity impact loading. *Composite Structures, 126*, 233–245.

65. Yoo, D. Y., Banthia, N., & Yoon, Y. S. (2016). Flexural behavior of ultra-high-performance fiber-reinforced concrete beams reinforced with GFRP and steel rebars.*Engineering Structures, 111*, 246–262.

66. Yoo, D. Y., Kang, S. T., & Yoon, Y. S. (2014a). Effect of fiber length and placement method on flexural behavior, tension-softening curve, and fiber distribution characteristics of UHPFRC. *Construction and Building Materials, 64*, 67–81.

67. Yoo, D. Y., Kwon, K. Y., Park, J. J., & Yoon, Y. S. (2015b). Local bond-slip response of GFRP rebar in ultra-high-performance fiber-reinforced concrete. *Composite Structures, 120*, 53–64.

68. Yoo, D. Y., Park, J. J., Kim, S. W., & Yoon, Y. S. (2014b). Influence of reinforcing bar type on autogenous shrinkage stress and bond behavior of ultra high performance fiber reinforced concrete. *Cement and Concrete Composites, 48*, 150–161.

69. Yoo, D. Y., Shin, H. O., Yang, J. M., & Yoon, Y. S. (2014c). Material and bond properties of ultra high performance fiber reinforced concrete with micro steel fibers.*Composites Part B: Engineering, 58*, 122–133.

70. Yoo, D. Y., & Yoon, Y. S. (2015). Structural performance of ultra-high-performance concrete beams with different steel fibers. *Engineering Structures, 102*, 409–423.

71. Yoo, D. Y., Yoon, Y. S., & Banthia, N. (2015c). Impact and residual capacities of ultra-high-performance concrete beams with steel rebars. In *Proceedings of the fifth international workshop on performance, protection & strengthening of structures under extreme loading*, East Lansing, MI.

72. Yoon, Y. S., Yang, J. M., Min, K. H., & Shin, H. O. (2011). Flexural strength and deflection characteristics of high-strength concrete beams with hybrid FRP and steel rebar reinforcement. In *Proceedings of the 10th symposium on fiber reinforced polymer reinforcement for concrete structures (FRPRCS-10), SP-275-04*, American Concrete Institute, Farmington Hills, MI, pp. 1–22.

73. Yudenfreund, M., Odler, I., & Brunauer, S. (1972). Hardened portland cement pastes of low porosity I. Materials and experimental methods. *Cement and Concrete Research, 2*(3), 313–330

5

CHAPTER

Engineered bamboo for structural applications

Bhavna Sharma, Ana Gatóo, Maximilian Bock, Michael Ramage

Department of Architecture, University of Cambridge, Cambridge, UK

HIGHLIGHTS

- Relevance of engineered bamboo products to construction and industry.
- Mechanical characterisation of bamboo scrimber and laminated bamboo is presented.
- Bamboo scrimber is shown to have similar strength to laminated bamboo.
- Engineered bamboo products have properties that are equal to or surpass that of timber.

ABSTRACT

Bamboo is a rapidly renewable material that has many applications in construction. Engineered bamboo products result from processing the raw bamboo culm into a laminated composite, similar to glue-laminated timber products. These products allow the material to be used in standardised sections and have less inherent variability than the natural material. The present work investigates the mechanical properties of two types of commercially available products – bamboo scrimber and laminated bamboo sheets – and compares these to timber and engineered timber products. It is shown that engineered bamboo products have properties that are comparable to or surpass that of timber and timber-based products. Potential limitations to use in structural design are also discussed. The study contributes to a growing body of research on engineered bamboo and presents areas in which further investigation is needed.

Keywords: Bamboo scrimber; Engineered composites; Laminated bamboo; Construction

1. INTRODUCTION

Bamboo has many advantages as a construction material: it is a rapidly renewable sustainable resource and has mechanical properties similar to timber. Worldwide, there is a growing interest in the development of bamboo products as a sustainable, cost-effective and ecologically responsible alternative construction material [1]. Partially due to the faster growth rate, and therefore harvest cycle, bamboo forests have up to four times the carbon density per hectare of spruce forests over the long term [2]. Bamboo is found in rapidly developing areas of the world where often timber resources are limited [2]. While the potential of bamboo is promising, more widespread development and use of bamboo is hampered by the lack of engineering data for mechanical properties and appropriate building codes [3] and [4].

Bamboo is an anisotropic material, having mechanical properties that vary in the longitudinal, radial and transverse directions. The raw material is a giant grass consisting of a hollow culm having longitudinal fibres aligned within a lignin matrix, divided by nodes (solid diaphragms) along the culm length (Figure 1). The thickness of the culm wall tapers from the base of the culm to the top. As a functionally graded material, the bamboo fibres also vary within the culm wall decreasing in density from the exterior to the interior (Figure 1).

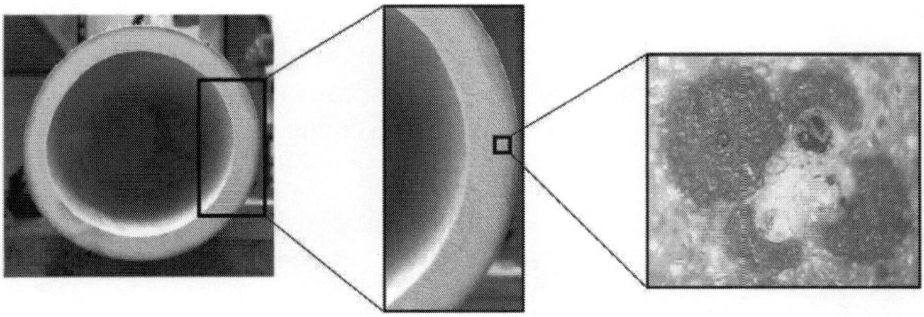

Figure 1. Details of a bamboo culm.

While there are more than 1200 species worldwide, full culm bamboo construction is limited by the variation in geometric and mechanical properties. The difficulty in making connections and joints suitable for round (and variable) sections is also prohibitive for mainstream construction; however increasing research demonstrates a growing industry and demand for sustainable building products. Studies vary from the use of full culm bamboo in construction and scaffolding (e.g., [5], [6], [7], [8], [9] and [10] to engineered bamboo composites (e.g., [11], [12], [13], [14], [15], [16], [17], [18], [19], [20] and [21]. Engineered bamboo composites are of particular interest due to the standardisation of shape and the relatively low variability in material properties [22].

Two examples of engineered bamboo are bamboo scrimber and laminated bamboo [22]. Bamboo scrimber, also referred to as strand woven or parallel strand bamboo, consists of crushed fibre bundles saturated in resin and compressed into a dense block (Figure 2). The process is materially efficient, utilising approximately 80% of raw inputs [23], and produces a product with a Janka hardness that is acceptable for external applications such as deck flooring. The process maintains the longitudinal direction of the bamboo fibres and utilises the resin matrix to connect the fibre bundles. In contrast, laminated bamboo maintains both the longitudinal fibres as well as a portion of the original culm matrix. The bamboo culm is split, planed, processed (bleached or caramelised), laminated and pressed to form the board product (Figure 3). The orientation of the strip within the board, and therefore the direction of the radial fibre density, is randomly placed within in the board (Figure 3). The final products use only approximately 30% of raw material input due to large losses of material when the strips are planed to form the rectangular section [23]. The sheet product is primarily used indoors for surface applications or furniture. While both materials are currently used for surface applications, both maintain the inherent strength of bamboo by maintaining the longitudinal fibre orientation and the engineered product creates a uniform section for connections and joints in structural applications.

Figure 2. Bamboo scrimber general manufacturing process in China.

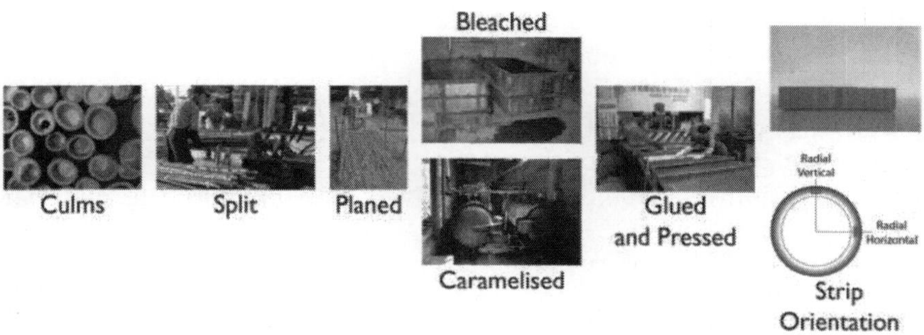

Figure 3. Laminated bamboo general manufacturing process in China.

The present work investigates the mechanical properties of bamboo scrimber and laminated bamboo to assess the potential for structural applications. A comparison with timber and engineered timber products is also presented.

2. MATERIALS AND METHODS

Two commercially produced products from China were used in the study. The bamboo scrimber product is comprised of *Phyllostachys pubescens* (Moso) with a phenol formaldehyde resin. The final product is a 140 × 140 mm section available in varying lengths. As shown in Figure 2 and discussed in the previous section, the process of manufacturing bamboo scrimber uses the bamboo culm with minimal processing. The resulting commercial product is tested as a final product with no additional modifications. The average density of the bamboo scrimber is 1160 kg/m^3 with a moisture content of 7%. In comparison, Moso as a raw material has a relative density of approximately 0.5–1.0.

Laminated bamboo sheets are also manufactured from Moso bamboo strips using a soy-based resin as shown in Figure 3 and discussed in the previous section. The structural specimens are built up from a commercial sheet (2440 × 1220 × 19 mm). The sheet was cut and the section laminated into the desired dimensions using polyurethane adhesive (Purbond HB S309). The adhesive was applied manually with a glue proportion of approximately 180 g/m^2 (final product) and the lamina pressed using manual clamps to apply the required pressure of 0.6 MPa for 4 hours (Figure 4). Two orientations were tested: radial horizontal and radial vertical, which refer to the orientation of the original strip within the beam, as shown in Figure 3. The industry terminologies for the equivalent orientations are edgewise and flatwise respectively, which refer to the orientation of the strips within a board, however this terminology is not well-defined for beam sections. Therefore the two orientations are herein described as radial horizontal and radial vertical and are shown inFigure 4. The laminated bamboo has an average density of 686 kg/m^3 with a moisture content of 6%. All specimens are conditioned in a constant temperature of 23 °C and a relative humidity of 55% for 2 weeks before testing.

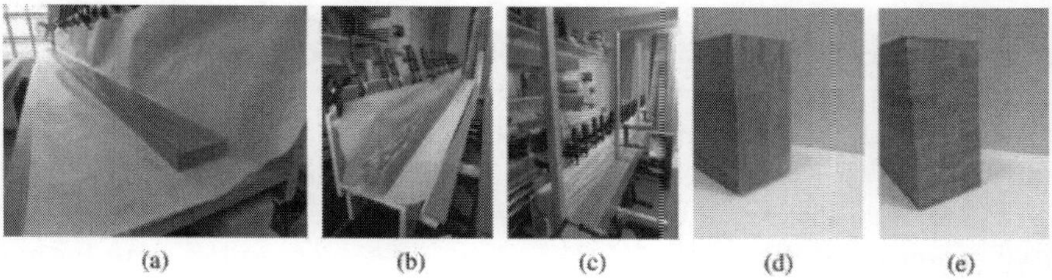

Figure 4. Laminated bamboo manufacturing process: (a) single lamina, (b) application of glue, (c) clamped specimen, (d) radial horizontal orientation, and (e) radial vertical orientation.

2.1. Standard test methods used

To obtain the material and mechanical properties, tension, compression, shear and flexural tests were conducted based on BS 373 *Methods of testing small clear specimens of timber* [24], ASTM D143 Standard *Test Methods for Small Clear Specimens of Timber*[25] and BS EN 408 *Timber structures - Structural timber and glue-laminated timber* [26]. BS 373:1957 cross references ASTM D143–09 and both are based on small clear specimens, whereas BS EN 408:2010 focuses on dimensional lumber samples. The standards, test methods, and test parameters used are summarised in Table 1.

Table 1. Experimental test methods for bamboo scrimber and laminated bamboo.

Standard	Test method	Test schematic	Direction	*n*	Specimen size (mm)	Loading rate (mm/min)
ASTM D143 [25]	Tension	a	Parallel to grain	10	25 × 25 × 460	1.0
		b	Perpendicular to grain	10	62 × 50 × 50	2.5
BS 373 [24]	Compression	c	Parallel to grain	10	20 × 20 × 60	0.635
		d	Perpendicular to grain	10	50 × 50 × 50	0.635
BS 373 [24]	Shear	e	Parallel to grain	10	50 × 50 × 50	0.635
BS EN 408 [26]	Four-point bending	f	Bamboo scrimber	10	40 × 40 × 800	6.35
			Laminated bamboo		60 × 120 × 2400	10

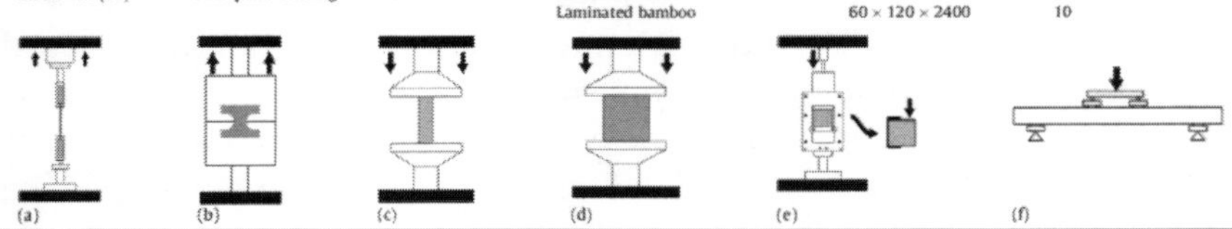

Both materials were tested using the same method with the exception of the bending test. BS EN 408:2010 for four-point bending defines the specimen length and loading rate as a function of the cross-sectional size. Due to the size limitations of the bamboo scrimber, it was necessary to test different dimensions and loading rate as given in Table 1. For all other mechanical tests, the small clear specimens were obtained from the beams at random and cut to the specified size [24] and [25]. For the tension parallel to grain, the specimens have two profiles specified in ASTM D143 [25]. They were initially cut with a water-jet and then routed to complete the second profile. Tension, compression and shear tests were conducted in a universal testing machine. All displacements are measured from the crosshead travel of test frame and the compliance of the frame was taken into account.

3. EXPERIMENTAL RESULTS

A summary of the test results is presented in Table 2 with additional observations for each mechanical property presented in the sections below. All of the mechanical properties were calculated in the linear elastic region at the limit of proportionality as required by the standards. Table 2 lists the average ($\bar{X}$) and the coefficient of variation (COV); ten specimens were tested in every case (i.e., $n = 10$).

Table 2. Material properties for structural bamboo and comparable natural bamboo and timber products.

		Density	Compression		Tension		Shear	Flexural		$\frac{f_b}{\rho}$
		ρ kg/m^3	$f_{c\parallel}$ MPa	$f_{c\perp}$ MPa	$f_{t\parallel}$ MPa	$f_{t\perp}$ MPa	$\tau_{\parallel}$ MPa	f_b MPa	E_b GPa	10^6 (m^2 s^{-2})
Laminated bamboo[a]	$\bar{x}$	686	77	22	90	2	16	77–83	11–13	16–19
	COV	0.05	0.05	0.07	0.26	0.13	0.05	0.06–0.08	0.05–0.06	
Bamboo scrimber[a]	$\bar{x}$	1163	86	37	120	3	15	119	13	11
	COV	0.02	0.02	0.05	0.14	0.13	0.11	0.08	0.04	
Raw Bamboo *Phyllostachys pubescens* [b,c]	$\bar{x}$	666	53	–	153	–	16	135	9	14
Sitka spruce [d,e]	$\bar{x}$	383	36	–	59	–	9	67	8	21
Douglas-fir LVL[f,g]	$\bar{x}$	520	57	–	49	–	11	68	13	25

[a] Present study.

[b] Ghavami and Marinho [27].

[c] de Vos [28].

[d] Lavers [29].

[e] Kretschmann [30].

[f] Kretschmann et al. [31].

[g] Clouston et al. [32].

3.1. Tension

Figure 5 shows a comparison of representative failures for both parallel to grain and perpendicular to grain tests. In tension parallel to grain ($f_{t\parallel}$), both the bamboo scrimber and the laminated bamboo exhibited linear behaviour prior to failure. The tension perpendicular to grain ($f_{t\perp}$) results emphasises the low strength of the bamboo scrimber and the laminated bamboo perpendicular to the fibre direction. The majority of the failures occurred at various locations within the anvil (Figure 5) and through the bamboo rather than within the adhesive for both materials. The parallel tensile strength was more than forty times the perpendicular strength.

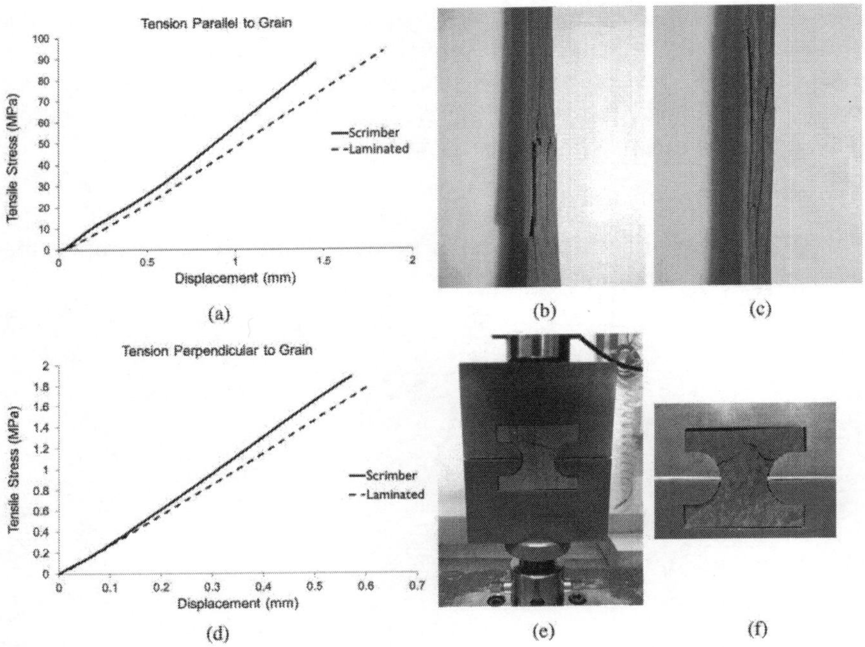

Figure 5. Tension parallel to grain (a) results; specimen failures in (b) laminated bamboo and (c) bamboo scrimber. Tension perpendicular to grain (d) results; specimen failures in (e) laminated bamboo and (f) bamboo scrimber.

3.2. Compression

The compression parallel to grain ($f_{c\parallel}$) results for both materials are shown in Figure 6a and in Table 2. The failure mode of the bamboo scrimber was overall section buckling of the specimen (Figure 6). A similar failure was noted by Huang et al. [14] in larger (105 × 105 × 315 mm) bamboo scrimber specimens. All of the laminated bamboo specimens failed in overall section buckling with no shear failures observed. Li et al. [33]also noted the buckling failure in larger samples (100 × 100 × 300 mm) in compression parallel to grain specimens that were manufactured from the upper portion of the bamboo with laminates of approximately 4 mm in thickness, similar to the specimens tested here (~4–6 mm). The representative curves show that both materials exhibit bilinear behaviour, which has also been observed in other studies [14] and [33]. Because specimen buckling dominates behaviour, the compressive strength is determined at the proportional limit in all cases (the knee in the bilinear curve). Thus this value must be interpreted as the strength for the sample dimension used rather than the compressive capacity of the material.

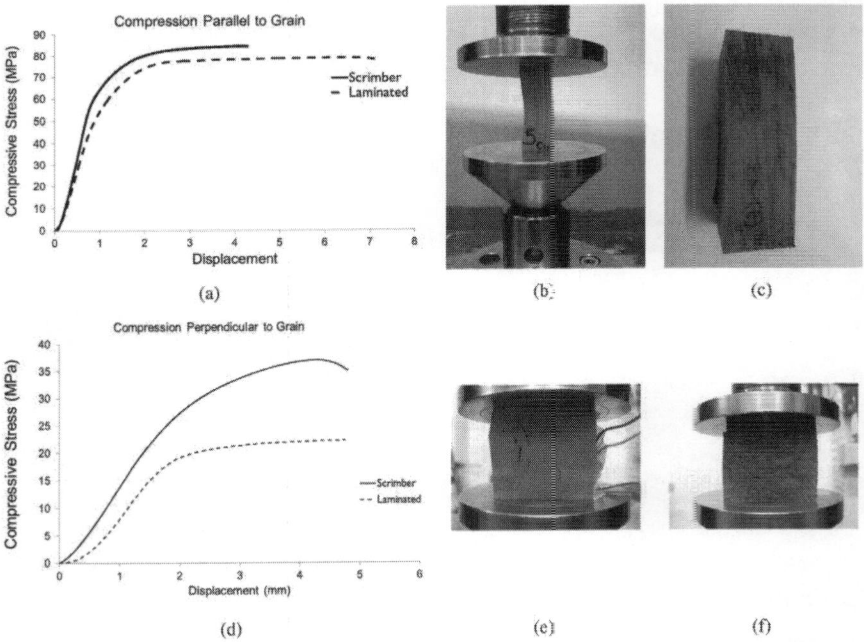

Figure 6. Compression parallel to grain (a) results, specimen failures in (b) laminated bamboo and (c) bamboo scrimber. Compression perpendicular to grain (d) results; specimen failures in (e) laminated bamboo and (f) bamboo scrimber.

The materials had a reduced strength in compression perpendicular to grain ($f_{t\perp}$), in comparison to the parallel to grain direction (see Table 2). Both displayed bilinear behaviour and the bamboo scrimber achieved twice the load at similar displacements before failure (Figure 6b). The failure mode for both materials was fracture of the matrix and bamboo fibres. For bamboo scrimber, the material demonstrated a slight influence of the direction of the glue line (i.e. the direction of the initial manufacturing compression), showing a 5% increase in stress parallel to the glue line. In comparison, the laminated bamboo showed no influence of the direction of the glue line (Table 2).

3.3. Shear

In shear parallel to grain ($\tau_{||}$), the behaviour of the two materials is quite similar, however the laminated bamboo is able to withstand increased load and displacement before failure. The laminated bamboo is noted to have a higher compressibility before the shear failure occurred within the fibres. The failure surface of the bamboo scrimber was much rougher (see Figure 7). The average shear strength, however, was comparable with the laminated bamboo, slightly higher (16 MPa) than the bamboo scrimber (15 MPa).

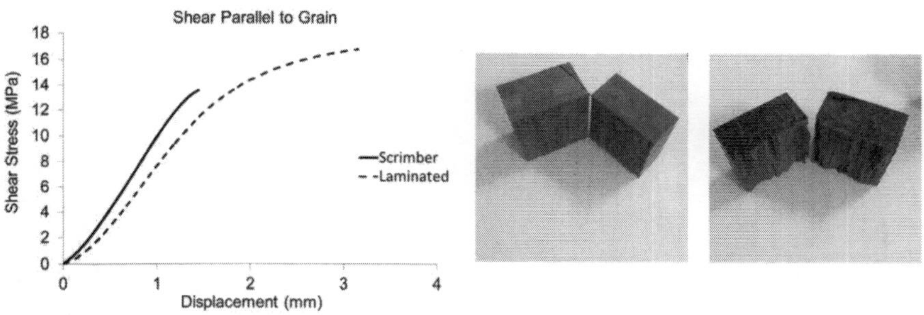

(a) (b) (c)

Figure 7. Shear Parallel to Grain (a) results; specimen failures in (b) laminated bamboo and (c) bamboo scrimber.

3.4. Bending

Bending results for the bamboo scrimber are shown in Figure 8a. The results indicated that a large displacement was achieved in the specimens before failure. Failure occurred at the tension faces of the specimens at mid-span. Similarly, the laminated bamboo failures occurred near mid-span, although these were characterised as longitudinal shear failures (i.e. VQ/I failures) within the depth of the beam for both the edgewise and flatwise oriented specimens (Figure 8b). All of the laminated bamboo specimens demonstrated brittle failure. Although micro-cracks within the material were audible and are also observed in small drops along the load–displacement curve (Figure 8b), no cracks were visible before failure. Comparison of the two orientations indicates that while achieving similar maximum loads, the flatwise orientation had an increased modulus of elasticity with a gain of approximately 18% over the edgewise orientation (Table 2).

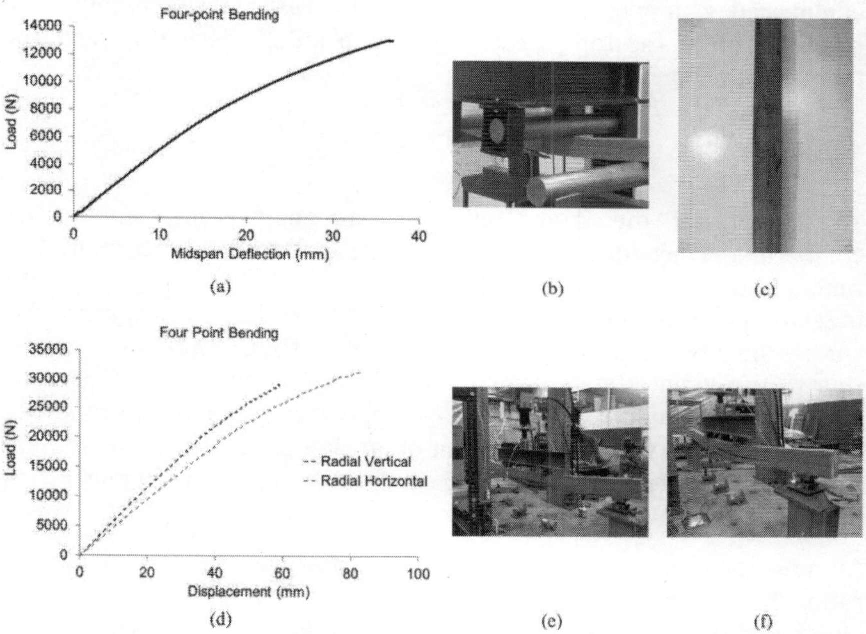

Figure 8. Four point bending (a) results and (b) specimen failure for bamboo scrimber. Laminated bamboo (c) results and specimen failures for (d) radial horizontal and (e) radial vertical orientations.

4. DISCUSSION

Both engineered bamboo materials, scrimber and laminated bamboo, exhibit the anisotropic behaviour typical of natural bamboo, and similar to that of fibre reinforced composites [34]. The mechanical behaviour of bamboo scrimber and laminated bamboo are very similar in tension, compression and shear parallel to grain. The laminated bamboo, however, has increased post-peak load deformation capacity, which is attributed to the compressibility of the matrix. Perpendicular to the fibre direction, the materials are also very similar, with the exception of compression where the bamboo scrimber has approximately twice the compressive strength of the laminated bamboo (Table 2). Overall, bamboo scrimber has slightly higher strengths in all properties with the exception of shear parallel to grain. The increased capacity is attributed to the densified fibres within the phenol formaldehyde resin. In flexure, engineered bamboo has the capacity to resist large amounts of compression at the top of the beam, with failure occurring near the tension face at mid-span.

4.1. Comparison to timber

The results of the study indicate that engineered bamboo has comparable mechanical properties to other structural materials like timber and raw bamboo. Table 2 provides a comparison of the presented results to mechanical properties from different experimental studies on timber, laminated veneer lumber (LVL) and raw bamboo. Due to the longitudinal strength of the bamboo fibres, both bamboo scrimber and laminated bamboo have increased strength in properties parallel to grain, with the exception of tension parallel to grain. Similar to timber, the strength perpendicular to grain is significantly lower than the strength parallel to grain.

An advantage of laminated bamboo is its flexural strength to density ratio. The specific bending modulus (E_b/ρ) is shown in Table 2. Laminated bamboo has a similar specific modulus (16–19 (10^6 m^2 s^{-2})) to Sitka spruce (21 (10^6 m^2 s^{-2})) and Douglas-fir LVL (25 (10^6 m^2 s^{-2})) and surpasses that of bamboo scrimber (11 (10^6 m^2 s^{-2})). Further comparison of the bending properties for various construction materials to engineered bamboo composites is shown in Figure 9. The graph illustrates that the flexural strength of engineered bamboo is on the higher end of the natural composite envelope and lower than fibre reinforced polymers composites. The failure mode of engineered bamboo must also be considered in comparison to other materials. Additional research is needed to understand the failure mode of the material and in particular the shear strength proportional to longitudinal capacity.

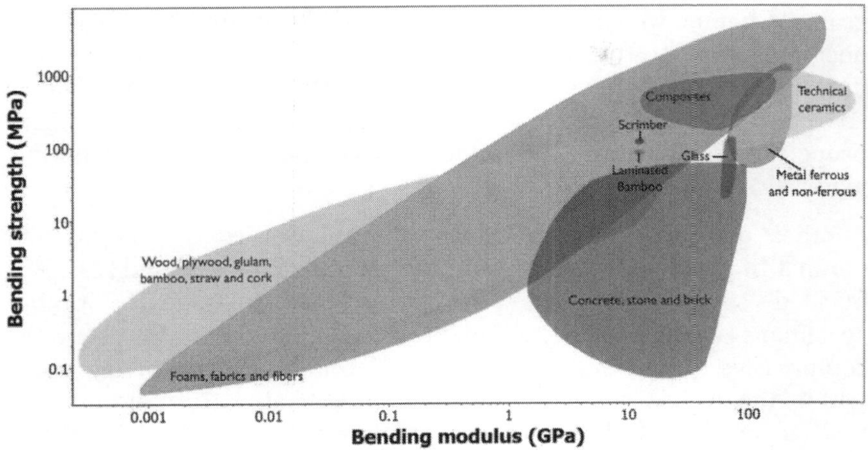

Figure 9. Bending modulus vs. bending strength for various construction materials [35].

4.2. Timber test methods

Timber standards are increasingly used in bamboo research and industry to characterise the material. While the observed failure modes and strengths appear to be similar to timber, further research is needed to understand the influences and applicability of the timber-based test methods for full-culm or engineered bamboo. For example, the BS EN 384: *Structural timber – Determination of characteristic values of mechanical properties and density* [36], applies factors to determine characteristic values for structural timber. The standard mechanical properties are defined for 12% moisture content with factors applied to the different properties to adjust for differences. A Chinese standard, *Testing methods for physical and mechanical properties of bamboo used in building,* also adjusts for 12% moisture content in full culm bamboo [37]. Engineered bamboo is at equilibrium at about 6% moisture content, well below 12% typical in timber and full-culm bamboo, and the impact on the mechanical properties has yet to be fully established. With increased moisture content above the material saturation point (MSP), timber demonstrates dimensional instability and decreases in strength. For example, the compressive strength increases rapidly as the moisture content decreases from the MSP [38]. Similarly, Wang et al. [39] and Xu et al. [40] explored the effect of moisture content and density on the compressive strength parallel to grain of bamboo elements and found that the compressive strength decreases rapidly until the material reached the saturation point. As natural materials, the impact of the uncertainty in the material is reflected in design factors, which can reduce the allowable strength properties significantly and affect the comparison to conventional materials.

Another example is the buckling failure observed in compression parallel to grain tests in this and other studies [12], [33] and [14]. The buckled shape is not an acceptable failure mode in timber specimens [24] and as discussed previously, does not represent the compressive strength of the material. The failure mode of the fibres within the matrix need to be further investigated to determine the mechanism at which failure in compression occurs and the relationship to the flexural buckling strength of the material.

To fully standardise engineered bamboo products, the test methods used to characterise the material must be evaluated to determine any influences or effects on the results. For example, bamboo scrimber and laminated bamboo in bending have similar properties to timber and glulam, however additional testing is needed to determine whether there are any bamboo-based factors that need to be applied to calculate appropriate characteristic values.

5. SUMMARY

The present work characterises the mechanical properties of two types of commercial products: bamboo scrimber and laminated bamboo. The study utilised timber standards for characterisation, which allows for comparison to timber and engineered timber products. The results of the study indicate that both products have properties that compare with or surpass that of timber.

Bamboo scrimber and laminated bamboo are heavily processed before testing. Future work includes investigating the influence of processing on the material properties. In particular, the impact of heat treatment performed on the material to achieve a caramel colour. A comparison study on natural coloured bamboo will provide better understanding of the effects of heat treatment on the strength of the material. The beam section can be optimised to take advantage of the high flexural strength to density ratio. Research on the influence of the orientation of the original board on the stiffness will also allow for further optimisation.

Further investigation of the influence of moisture and the density on the mechanical properties is needed to provide a foundation from which to develop design characterisation factors for engineered bamboo. Additional testing of full-scale specimens would also elucidate any effects in comparison to small clear specimens, as well as allow further comparison to timber and provide an additional step forward towards construction.

ACKNOWLEDGEMENTS

The presented work is supported by EPRSC Grant EP/K023403/1 and the Newton Trust, and forms part of a collaboration between the University of Cambridge, Massachusetts Institute of Technology (MIT) and University of British Columbia (UBC).

REFERENCES

1. De Flander K. The role of bamboo in global modernity: from traditional to innovative construction material. Thesis: Wageningen University, Wageningen Netherlands; 2005.
2. Yiping L et al. Technical Report 32: Bamboo and Climate Change Mitigation. Beijing, China: INBAR; 2010.
3. Harries KA, Sharma B, Richard MJ. Structural use of full culm bamboo: the path to standardization. Int J Archit Eng Constr 2012;1(2):66–75.
4. Gatóo A, Sharma B, Bock M, Mulligan H, Ramage M. Sustainable structures: bamboo standards and building codes. Proc ICE Eng Sustainability 2014;167(5):189–96.
5. Chung KF, Yu WK. Mechanical properties of structural bamboo for bamboo Scaffoldings. Eng Struct 2002;24:429–42.
6. Kaminsky S. Engineered bamboo houses for low-income communities. Struct Eng 2013;91(10):14–23.
7. Muthukaruppan R. Hong Kong – bastion of bamboo scaffolding. Proc ICE Civ Eng 2008;161(4):177–83.

8. Trujillo DJA, Ramage M, Change W. Lightly modified bamboo for structural applications. Proc Inst Civ Eng Constr Mater 2013;166(4):238–47.

9. Yu WK, Chung KF, Chan SL. Column buckling of structural bamboo. Eng Struct 2003;25:755–68.

10. Yu WK, Chung KF, Chan SL. Axial buckling of bamboo columns in bamboo scaffolds. Eng Struct 2005;27:61–73.

11. Bansal AK, Prasad TRN. Manufacturing laminates from sympodial bamboo – an Indian experience. J Bamboo Rattan 2004;3(1):13–22.

12. Correal JF, Lopez LF. Mechanical properties of Colombian glued laminated bamboo. In: Mod Bamboo Struct. London, UK: CRC Press Taylor & Francis Group; 2008. p. 121–8.

13. Correal J, Ramirez F, Gonzalez S, Camacho J. Structural Behavior of Glued Laminated Guadua Bamboo as a Construction Material. In: Proceedings of the 11th World Conference on Timber Engineering. Italy: Trentino; 2010.

14. Huang D, Zhou A, Bian Y. Experimental and analytical study on the nonlinear bending of parallel strand bamboo beams. Constr Build Mater 2013;44:592–5.

15. Lee AWC, Bai X, Bangi AP. Selected properties of laboratory-made laminatedbamboo lumber. Holzforschung 1998;52:207–10.

16. Mahdavi M, Clouston PL, Arwade SR. A low-technology approach toward fabrication of laminated bamboo lumber. Constr Build Mater 2012; 29(4):257–62.

17. Pereira MA, Faria OB. Bambu Project: Mechanical Characteristics of the Glued Laminated Bamboo. Thailand: In Proceedings from the VIII World Bamboo Congress; 2009. pp. 135–150.

18. Sinha A, Way D, Mlasko S. Structural performance of glued laminated bamboo beams. J Struct Eng 2014;140(1).

19. Sulastiningsih IM, Nurwati. Physical and mechanical properties of laminated bamboo board. J Trop For Sci 2009;21(3):246–51.

20. Xiao Y, Yang RZ, Shan B. Production, environmental impact and mechanical properties of glubam. Constr Build Mater 2013;44:765–73.

21. Zhou A, Bian Y. Experimental study on the flexural performance of parallel strand bamboo beams. Sci World J 2014.

22. Sharma B, Gatoo A, Bock M, Mulligan H, Ramage M. Engineered bamboo: state of the art. Proc ICE Constr Mater 2014.

23. van der Lugt P. Design Interventions for Stimulating Bamboo Commercialization: Dutch Design meets Bamboo as a Replicable Model. Delft, Netherlands: VSSD; 2008.

24. BS. British Standard 373–1957: methods of testing small clear specimens of timber. London, UK: British Standards Institution; 1957.

25. ASTM. ASTM D143–09: standard test methods for small clear specimens of timber. West Conshohocken, PA, USA: ASTM International; 2009.

26. BS. British standard EN 408–2010: timber structures – structural timber and glue-laminated timber. London, UK: British Standards Institution; 2010.

27. Ghavami K and AB Marinho (2001). Determinição das propriedades mecanicas dos bambus das especies: moso, matake, Guadua angustifolia, guadua tagoara e Dendrocalamus giganteus, para utilização na engenharia. Rio de Janeiro: PUC-Rio, Publicação RMNC-1 Bambu 01/2001 (in Portuguese).

28. de Vos V. Bamboo for exterior joinery: a research in material properties and market perspectives. Thesis: Larenstein University, Leeuwarden, Netherlands; 2010.

29. Lavers GM (2002) The strength properties of timber, 3rd edition. Building Research Establishment (BRE) Report Series, London, UK.

30. Kretschmann, DE (2010) Mechanical Properties of Wood. In Wood Handbook, General Technical Report FPL-GTR-190. U.S. Department of Agriculture, Forest Service, Forest Products Laboratory, Madison, WI, Chapter 5, pp. 5–1 - 5–46.

31. Kretschmann DE, Moody RC, Pellerin RF, et al. (1993) Effect of various proportions of juvenile wood on laminated veneer lumber. Technical Report FPL-RP-521. U.S. Department of

Agriculture, Forest Service, Forest Products Laboratory, Madison, WI.

32. Clouston P, Lam F, Barret JD. Incorporating size effects in the Tasi-Wu strength theory for Douglas-fir laminated veneer. Wood Sci Technol 1998; 32(3):215–26.

33. Li H, Zhang Q, Huang D, Deeks AJ. Compressive performance of laminated bamboo. Compos B 2013; 54:319–28.

34. Amada S, Untao S. Fracture properties of bamboo. Compos B Eng 2001; 32(5):451–9.

35. CES. Cambridge engineering selector edupack software. Cambridge, UK: Granta Design Limited; 2013.

36. BS. British standard EN 384–2010: structural timber – determination of characteristic values of mechanical properties and density. London, UK: British Standards Institution; 2010.

37. CSA. Chinese Standards Agency JG/T199: testing methods for physical and mechanical properties of bamboo used in building. Beijing, China: Chinese Standards Agency; 2007.

38. Dinwoodie JM. Timber – a review of the structure – mechanical property relationship. J Microsc 1975;104(1):3–32.

39. Wang H, Li W, Ren D. A two-variable model for predicting the effects of moisture content and density on compressive strength parallel to the grain for moso bamboo. J Wood Sci 2014.

40. Xu Q, Harries KA, Li X, Lui Q, Gottron J. Mechanical properties of structural bamboo following immersion in water. Eng Struct 2014;81:230–9.

6

CHAPTER

Effect of Air Entrainment on Shrinkage of Blended Cements Concretes

Wojciech Piasta[a], Hubert Sikora[b]

[a] *Faculty of Civil Engineering and Architecture, Kielce University of Technology, Tysiaclecia PP 7, 25-314 Kielce, Poland*

[b] *GAMEX-PROJEKT Co. Ltd., Paderewskiego 31/47, 25-004 Kielce, Poland*

HIGHLIGHTS

- Shrinkage of air entrained (AE) concretes was higher than that of non AE concretes.
- Increase in shrinkage of concretes with increasing content of air voids was observed.
- Higher shrinkage might be due to air voids and porous air void–paste interface.
- Interconnection and overlapping of interfaces increased vapour diffusion and shrinkage.
- Bigger volume of cement paste in AE concrete may also increase shrinkage strains.

ABSTRACT

Shrinkage strains of air-entrained (AE) concretes were bigger than strains of non-air-entrained concretes made with the same cements. The results clearly showed an increase in strains with increasing content of air voids, regardless of cement type. Higher shrinkage strains of AE concretes may be most probably related to their microstructure of complex porosity and in particular, to porous cement paste adjacent to air void. Occurrence of air void–cement paste interfacial transition zones of higher porosity as well as their overlapping and interconnection may increase vapour diffusion accelerating moisture loss and drying shrinkage. Another likely reason for shrinkage increase may be increase in the volume of cement paste and decrease in the volume of aggregate in AE concrete.

Keywords: Concrete; Shrinkage; Air entrainment; Air content; Blast furnace slag; Limestone powder

1. INTRODUCTION

Shrinkage is a complex process which involves a lot of summatively overlapping factors. Shrinkage of cement paste is a result of a decrease in its volume during the hydration and drying of cement. During cement hydration, water and cement phases are substrates whose volume is bigger than that of the hydration products. This decrease in volume is observed before and during cement setting and is known as a chemical shrinkage or contraction [1]. The primary driving force for shrinkage is capillary tension in the pores, occurrence of disjoining pressure and changes in the tension of solid gel molecules [2]. Concrete shrinkage is also influenced by the aging of cement paste. The polymerization reactions that constitute aging increase the stiffness (modulus) of the paste. The monomeric molecule of C—S—H liberates a water molecule whereby Si—O—Si bonds are formed. This mechanism of condensation densifies the gel and it explains why young (unaged) pastes exhibit irreversible shrinkage [3].

Irrespective of chemical and plastic shrinkage of early-age concrete, the total shrinkage may be, in the simplest way, divided into autogenous shrinkage and drying shrinkage. Autogenous shrinkage is uniform and isotropic and takes place in the entire volume. When concrete stiffens due to hardening of the cement paste, autogenous shrinkage is initiated [2]. The unhydrated cement absorbs water from capillary pores and nanopores for further hydration. Then, the self-desiccation follows and autogenous shrinkage occurs which in turn develops until hydration stops [4]. Unlike ordinary concretes, low w/c high performance concretes (HPC) made with bulk amount of cement show much more autogenous shrinkage due to shortage of capillary water necessary for cement hydration.

Replacement of cement with limestone powder (LP) and fly ash (FA) may decrease total shrinkage and its rate and, as a result, reduce the probability of short-time concrete cracking [5]. It was also observed that drying shrinkage lessened with the use of blast furnace slag (BFS), FA and metakaolinite [6]. However, there is no consensus that replacing part of cement with BFS decreases every type of shrinkage [7]. The results demonstrated that partial replacement of 50% and 70% of white Portland cement with slag reduced the total shrinkage by as much as about 25% and 40%, respectively. The quantification separately of autogenous and drying shrinkage of natural pozzolan, limestone and BFS cement mortar showed a very large – above 50% – decrease in autogenous shrinkage of mortar by replacement of 50% of cement with BFS [8]. Although BFS cement mortar showed higher drying shrinkage, its total shrinkage was still the smallest. Other test results showed that BFS cement concrete exhibited higher autogenous shrinkage than ordinary portland cement (OPC) concrete and that the higher the BFS content, the higher the autogenous shrinkage [9]. FA can reduce the shrinkage, and silica fume can increase the shrinkage, and the effect of BFS may be between the two mentioned [10]. Not all quoted test results on shrinkage of blended cement concretes are consistent. However, while evaluating the effect of mineral additives on shrinkage, it is important to take into account the specific surface of the additives and their relative content because a very large specific surface of the additive compared to OPC may increase shrinkage strain in the same way as the silica fume does. Questions arise, though, what the effect of air entrainment on shrinkage of blended cement concretes might be and whether the mineral additives could decrease shrinkage of AE concrete. It should be borne in mind that the more air-entraining agent (AEA) is needed for air-entraining blended cement concretes the larger mineral additive content (BFS, LP or FA) in concrete is. Moreover, it is more difficult to obtain proper features and arrangement of air voids in blended cement concretes compared to OPC [11].

Capillary porosity and capillary pore diameter distribution related to w/c ratio undeniably belong to main factors affecting shrinkage. However, while considering and investigating shrinkage of concrete containing air voids, one has to focus on the fact that they are connected with capillary pores and interconnected through capillary pores, too, which are the main path of moisture transport in the hardened paste of AE concrete. The entrained air voids can increase or decrease bulk transport properties depending on the transport mechanism under consideration [12]. Influence of air voids on transport properties suggests also possibility of their influence on shrinkage.

The drying shrinkage develops from the surface into concrete, just as moisture begins to leave the paste. The drying rate and shrinkage rate are not uniform since at the onset of drying their rate is relatively fast, but it gradually decreases with time given that under nearly saturated conditions both bulk water and vapour fluxes participate in moisture transport [13]. However, moisture movement in concrete in the form of vapour flux is dominant when pore-relative humidity is within the range of 15–95%. Then vapour flux is governed by vapour diffusion in unsaturated pore space [14]. That is why it is possible and favourable to describe moisture transport in concrete using diffusivity. Jafarifar et al. [15]determined that there exists a relationship between moisture diffusivity and moisture content of the unsaturated concretes and a relationship between free shrinkage and moisture loss. This, in turn, indicates that the shrinkage strains of unsaturated concrete are consistent with moisture diffusivity.

Since concrete contains 25–35% of cement paste in its volume and all air voids reside in cement paste, the volume of cement paste is usually increased by about 5–15% when, for instance, the entrained air content is equal to 3–6% of concrete volume. Thus, the occurrence of entrained air as a new, not insignificant fraction of cement paste volume may affect the shrinkage of concrete, all the more so that according to a known rule for ordinary and high performance concretes shrinkage is proportional to content of cement paste in concrete mix [16].

Due to air entrainment, the microstructure of hardened concrete is more heterogeneous because new fractions occur in it. Rashed and Williamson [17] have identified two distinct features to air voids: a shell to the air void surface and an interfacial transition zone between this shell and the bulk cement paste. The shell appears to be made up of small (1–5 μm) mineral particles [18]. Whereas, Ley et al. [19] showed that air void surface is made of CSH phase, with calcium/silicon ratio equal only to 1.1. For a bulk OPC paste, the ratio is 1.5. Due to the presence of water-rich fresh paste containing lower cement content around the air voids interfacial paste has higher initial *w/c* ratio compared to bulk paste farther away from the interface. The width of the interface zone adjacent to the shell is around 30 μm from its surface [12]. In general, it appears that the air void–paste interface is similar to aggregate–cement paste interfacial transition zone whose width is 20–50 μm. However, no precipitation of calcium hydroxide was observed at the air void–paste interface [18]. In extensive research of the microstructure in AE concretes and mass transport properties Wong et al. [12] stated that the porosity near the air void boundary is about 2–3 times that of the bulk paste and that air entrainment increases gaseous diffusivity and permeability by up to a factor of 2–3 in case of the highest air contents. Interfacial paste of porous microstructure occupies a large part of cement paste volume. The air void interfaces can overlap and interconnect. Therefore, the occurrence of air voids and related to them numerous changes in microstructure of AE concrete cannot be neutral for its properties.

Combining the conclusions based on AE concrete transport property research with shrinkage and concrete transport properties, certain relationships between drying shrinkage of AE concretes and vapour diffusivity can be observed. In AE concretes, gaseous diffusivity increases as the entrained air content grows. This, in turn, could suggest that air entrainment may result in accelerating vapour flux as well as affect water loss and shrinkage strains due to increased diffusivity. Moisture transport in AE concretes is more complex, as they have a wide variety of pore structures, permeable air voids and higher porosity of interfacial transition zone around air voids. Therefore, a certain effect of air entrainment on shrinkage strains of concretes is probable.

Though most shrinkage types of concrete containing a large variety of cements, aggregates and additives were thoroughly tested, a few observation and test results about the range of shrinkage of AE concrete have been published. According to scant data, the air entrainment of OPC concrete should not have any influence on its shrinkage. Fagerlund [20], when testing AE concrete properties, concluded that air entrainment does not affect shrinkage of OPC concretes. Also Keene [21], quoted by Neville [16], showed that concrete shrinkage is independent on air entrainment. There are in fact none for blended cement concretes. Therefore, an attempt was made to investigate the shrinkage of AE concretes manufactured from blended cements.

2. EXPERIMENTAL PROCEDURE

2.1. Aim and scope of the experiment

The aim of this study has been to investigate long-term shrinkage strains of air-entrained and non-air-entrained concretes to find the influence of entrained air voids. Concretes made with the use of several cements have been tested to determine whether mineral additives replacing part of OPC alter shrinkage strains of AE concretes as in the case of ordinary concretes. The aim has been also to identify the effect of increase in content of air voids on shrinkage strains of BFS cement and OPC concretes. Scanning electron microscopy (SEM) has been used to determine how the microstructure of air-entrained concrete can influence shrinkage strains. The effect of air entrainment on different strains has been also considered in relation to the content of cement paste in concrete containing various air volume.

2.2. Methods

The measurements of shrinkage strain were carried out with the Amsler's extensometer using 6 samples of each concrete in prisms 10 × 10 × 50 cm with pins embedded in concrete centrally in the front part. The first measurement of the samples was carried out 24 h after their casting. Five samples sized 15 × 15 × 15 cm of each concrete type cured for 28 days were used to test compressive strength. The total duration of shrinkage tests was up to 300 days. However, shrinkage strain measurements were carried out for 180 days in the additional tests. Samples have been cured under conditions of relative humidity (rh) of 55 ± 5% and at the temperature of 20 °C.

The earliest measurements of sample length were only possible after their demoulding, i.e. after 24 h from their casting. Therefore, shrinkage occurring during the first 24 h has not been recognised. The first shrinkage strains have been identified between 24 and 48 h from casting. Long term deformations shown in the figures are the total of drying shrinkage and part of autogenous shrinkage, which is not dominant in ordinary concretes. Therefore, the deformations do not represent the total shrinkage. Further on in this paper they are called shrinkage tested from day 1 (24 h) when they were measured for the first time.

In addition, microstructure of air-entrained OPC concretes and 50% BFS cement concretes were tested with a SEM with EDXA. While observing the microstructure, air void shells and air void–cement paste interfacial transition zones have been given special attention. The air-entrained concretes tested with SEM were cured for 180 days. For SEM tests the sample without coarse aggregate has been fractured from the concrete. A small flake from interior surface approx. 8–10 mm thick was stored in vacuum until it was examined with a SEM with EDXA. The SEM used was Environment SEM. The operating voltage was 20 kV, working distance 9.5–12 mm.

2.3. Materials

The samples for the tests have been cast from 22 concrete mixes, including 9 non-air-entrained (nAE) concrete mixes and 13 AE concrete mixes. In total, OPC as well as 4 blended cements and 3 types of coarse aggregate have been used for all the concretes.

To determine the combined influence of two factors i.e. air entrainment and mineral additives, on concrete shrinkage, two series of experiments have been carried out. The tests on shrinkage strain have been carried out for AE and nAE concretes with the use of 2 coarse aggregates, gravel and crushed carbonate aggregate, to ensure reliability of the results.

Shrinkage strain tests have been carried out for following concretes:

–8 (4 pairs) AE concretes and nAE concretes from blended cements: limestone Portland cement (CEM II/A-LL 42,5R – also marked as 14% LP), blast furnace slag Portland cement (CEM II/B-S 42,5N – also marked as 32% BFS), blast furnace slag cement (CEM III/A 42,5N – also marked as 55% BFS), and from OPC (CEM I 42,5R) and river gravel,

–6 (3 pairs) AE concretes and nAE concretes manufactured from coarse crushed carbonate aggregate and 2 slag blended cements (CEM II/B-S, CEM III/A) with various content of slag (32% BFS, 55% BFS) and from OPC (CEM I 42,5R) for comparison.

Moreover, two-way analysis of variance with replication was conducted and its results have been briefly presented in the form of plot (Figure 11) taken from analysis of variance in Discussion. When the variance analysis confirmed the significant differences between strains results of AE and nAE concrete, additional tests of concretes with various content of entrained air have been carried out to obtain a more detailed estimation of air entrainment influence on concrete shrinkage. The concretes tested have been manufactured from OPC or 50% BFS cement and basalt aggregate.

All five cements have been manufactured in the same cement plant and made from the same cement clinker. Phase composition of clinker is presented in Table 1.

The content of mineral additives in blended cements has been presented in the Table 2, whereas physical properties of cements in Table 3.

Table 1. Phase composition and alkali content in cement clinker (%).

Compound	C_3S	C_2S	C_3A	C_4AF	Free CaO	Na_2O_e
Content	59.50	19.10	8.90	9.10	0.80	0.65

Table 2. Content of mineral additives in blended cements (%).

Cement type	BFS	LP
CEM II/B-S (32% BFS)	32	4
CEM III/A (50% BFS)[a]	50	4
CEM III/A (55% BFS)	55	4
CEM II/A-LL (14% LP)	–	14

[a] 50% BFS cement used for basalt aggregate concretes to check an influence of air content.

Table 3. Physical properties of cements.

Cement	OPC	CEM II/A-LL 14% LP	CEM II/B-S 32% BFS	CEM III/A 50% BFS	CEM III/A 55% BFS
Le Chatelier, mm	0.4	0.4	0.4	0.4	0.4
Specific surface[a], cm^2/g	3570	4070	4050	4250	4410
Specific gravity, kg/dm^3	3,09	3.04	3.03	2.99	2.98
Initial setting time, min	182	179	235	235	245
Heat of hydration after 41 h[b], J/g	337	327	304	294	291
Compressive strength, MPa					
2 days	26.9	26.3	21.3	16.5	14.6
28 days	51.9	50.6	55.8	54.5	57.2

[a] Specific surface determined for cement containing additive.

[b] Heat of hydration measured with the use of semiadiabatic method.

In order to entrain air voids in concrete mixes, synthetic sodium alkyl sulphonate (AEA) was used in the amount of 0.15–0.31% of the cement mass (0.50–1.05 kg/m³). The coarse aggregates applied were river gravel and crushed carbonate and basalt aggregates of nominal maximal grain size 16 mm. The fine aggregate was river sand. The particle size gradation of the coarse and fine aggregates has been shown in Table 4.

In order to investigate the influence of air entrainment on shrinkage of blended cement concretes, it has been initially assumed that air content (A_m) in concrete mixes should be the same as the optimum amount for frost-proof concretes which is 5.5% ± 0.5%. Moreover, the influence of air content within the range of 2–8% on OPC and 50% BFS cement concretes has been tested. The proportions of concrete mixes have been presented in Table 5.

Table 4. Sieve analysis of fine and coarse aggregate, %.

Sieve size, mm	Sand	Coarse aggregate		
		Gravel	Carbonate	Basalt
31.5	100	100	100	100
16	100	98	96	98
8	100	46	44	48
4	100	9	10	11
2	92	2	1	3
1	68	1	0	1
0.5	37	0	0	0
0.25	11	0	0	0
0.125	2	0	0	0

Table 5. Proportions of concrete mixes (kg/m³).

Component	Concrete		
	Gravel	Carb. agg.	Basalt
Cement	350	350	350
Water	175	175	175
Coarse agg. 2–8 mm	560	580	630
Coarse agg. 8–16 mm	695	725	790
Quartz sand	620	620	620
w/*c*-ratio	0.50	0.50	0.50

The following tests of concrete mixes were conducted: air content with the pressure method, bulk density and slump. Air content and properties of concrete mixes have been presented in Table 6. It must be noted that the air content in blended cement mixes was equal 5.2–5.8%, as initially assumed. Compressive strength of concretes cured for 28 days has been shown in Table 6.

Table 6. Properties of mixes and 28-day compressive strength of concretes.

Cement	OPC		14% LP		32% BFS		55% BFS	
Air entrainment	nAE	AE	nAE	AE	nAE	AE	nAE	AE
Aggregate	Gravel							
Air content (%)	2.2	5.6	2.4	5.8	2.0	5.2	2.2	5.4
Slump (mm)	30	56	30	40	10	50	40	55
Bulk density (kg/m^3)	2350	2260	2340	2250	2360	2250	2340	2230
Vol. of paste (incl. air) (dm^3/m^3)	304	328	311	331	305	327	308	331
Vol. of aggregate (dm^3/m^3)	696	672	689	669	695	673	692	669
Compr. strength (MPa)	46.1	38.9	43.3	38.1	51.8	43.2	52.0	45.8
Aggregate	Crushed carbonate aggregate							
Air content (%)	2.1	5.4	–	–	2.0	5.7	2.1	5.2
Slump (mm)	32	53	–	–	23	50	35	43
Bulk density (kg/m^3)	2390	2310	–	–	2395	2320	2400	2310
Vol. of paste (incl. air) (dm^3/m^3)	303	326	–	–	305	331	307	329
Vol. of aggregate (dm^3/m^3)	697	674	–	–	695	669	693	671
Compr. strength (MPa)	41.9	35.1	–	–	42.5	35.7	41.0	35.2

BFS cement and OPC concretes, made to test air content influence on shrinkage, contained between 3.4% and 8.1% air. Properties of concretes and their mixes have been shown in Table 7.

Table 7. Properties of basalt aggregate concrete mixes with different air content.

Cement	50% BFS				OPC			
Air entrainment	nAE	AE			nAE	AE		
Air content in concrete mix (%)	2.0	3.7	5.2	7.8	2.1	3.4	5.4	8.1
Slump (mm)	30	50	65	85	40	60	75	90
Bulk density (kg/m^3)	2515	2470	2435	2370	2520	2490	2435	2365
Vol. of total air (dm^3/m^3)	20	37	52	78	21	34	54	81
Vol. of paste (dm^3/m^3)	306.3	318.4	329.0	347.4	303.5	316.6	327.0	344.7
Air content in paste (incl. air) (%)	6.5	11.6	15.8	22.5	6.9	10.7	16.5	22.9
Compr. strength (MPa)	50.6	46.6	42.2	36.0	48.5	45.5	41.0	35.1

3. TEST RESULTS

3.1. The influence of air entrainment on shrinkage strains of blended cements concretes

Figure 1 shows tests results for shrinkage strains of 14% LP cement concretes. The shrinkage of AE concrete throughout the whole experiment period was higher than the shrinkage of nAE one. The highest difference between strains of both concretes, amounting to almost 100 * 10^{-6}, appeared at the age between 7 and 60 days. At the end of the test, the difference was 70 * 10^{-6} (i.e. more than 10%), which can be assumed as statistically significant.

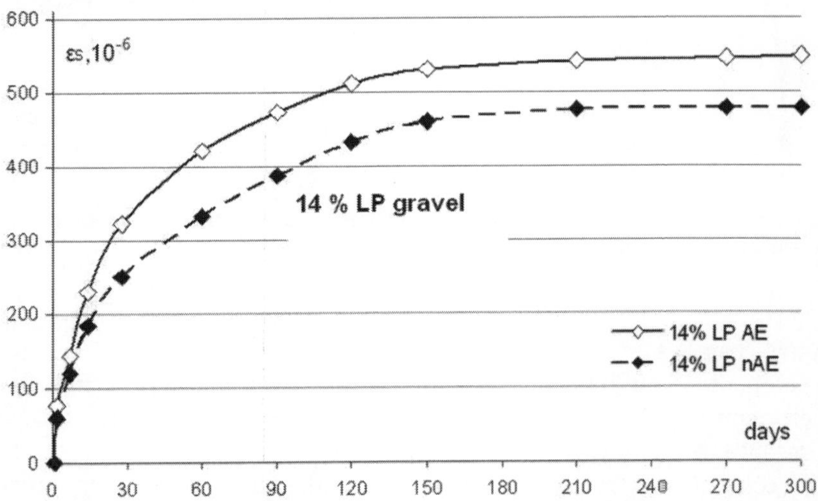

Figure 1. Shrinkage strains of AE and nAE concretes from cement containing 14% LP.

Figure 2 shows shrinkage strains of four 32% BFS cement concretes. The concretes have been manufactured from two coarse aggregates, gravel and carbonate aggregate both in case of AE and nAE concretes. When comparing the shrinkage strains of AE and nAE concretes in pairs i.e. strains of 2 gravel concretes and 2 concretes from carbonate aggregate, it can be noticed that shrinkage of AE concretes is larger than that of nAE ones in every case. The biggest difference in the strains induced by air entrainment appeared in both cases at the end of the experimental period. In the case of gravel concrete the difference at the age of 300 days amounted to about 120 * 10^{-6}, that is 33%.

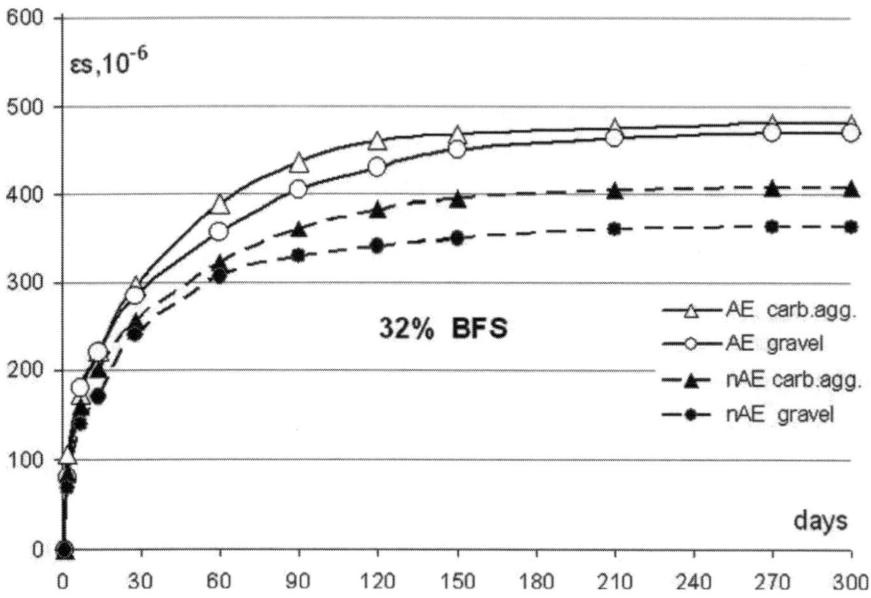

Figure 2. Shrinkage strains of AE and nAE concretes from carbonate aggregate or gravel and cement containing 32% BFS.

Figure 3 shows test results of shrinkage strains of both AE and nAE concretes made of BFS cement, containing 55% slag. The concretes have been manufactured from two coarse aggregates i.e. gravel or carbonate aggregate. A significant effect of air entrainment on shrinkage of 55% BFS cement concrete has been observed. Air entrainment caused increase in shrinkage of both gravel and carbonate aggregate concretes. The strains of AE concretes throughout the whole experiment period were larger than the strains of non-air-entrained ones, and the difference between them increased with time. At the age of 300 days, the difference in the strains amounted to 34% for the concretes from carbonate aggregate and 32% for gravel concretes.

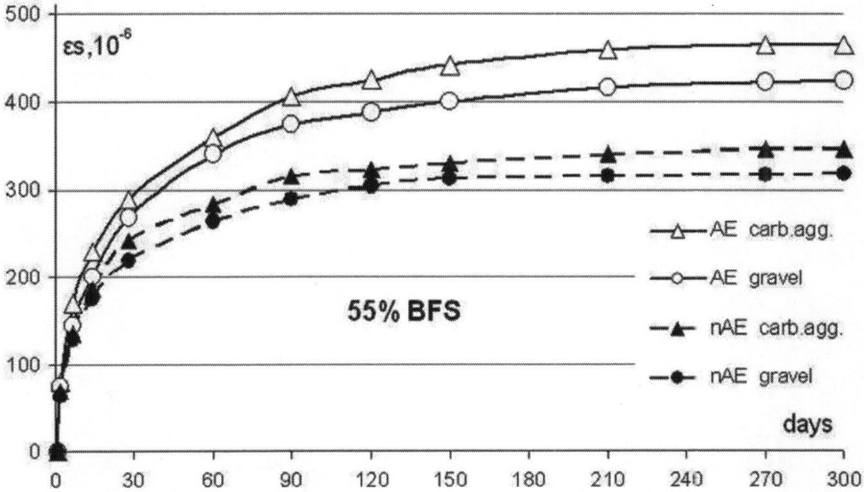

Figure 3. Shrinkage strains of AE and nAE concretes from carbonate aggregate or gravel and from cement containing 55% BFS.

3.2. Air content influence on concrete shrinkage

Test results on BFS cement concretes made with basalt aggregate (Figure 4) have proved that shrinkage strains increase along with an increase in the content of entrained air in the concrete mixes. Although there was no direct proportionality between shrinkage and entrained air content, shrinkage of 50% BFS cement concretes tended to increase along with an increase in entrained air-voids volume across the entire range of obtained values. Furthermore, the differences between final strains of individual concretes with various amounts of air voids are not low. When concretes were cured for 180 days the difference between shrinkage strain of concrete containing total air up to 7.8% of its volume and that of non-air-entrained concrete amounted to about $250 * 10^{-6}$ and exceeded 70%.

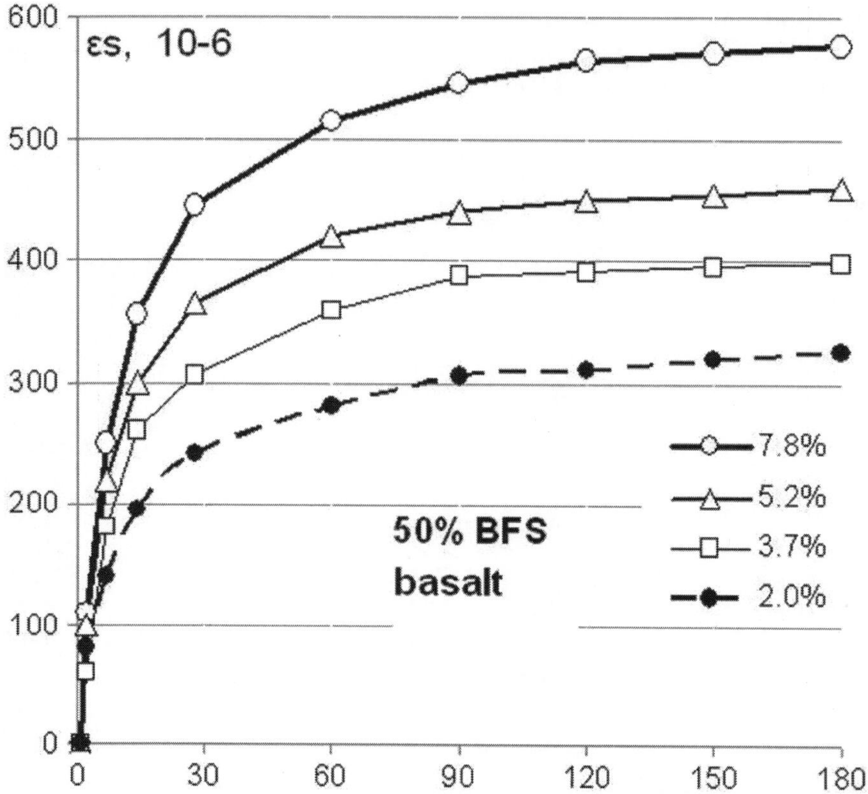

Figure 4. Influence of air content on shrinkage strains of 50% BFS cement concrete with basalt aggregate.

Moreover, the shrinkage strains of OPC concretes increased as the content of entrained air grew. However, the strains of OPC concretes (Figure 5) with various content of air developed in time in a slightly different way than in BFS cement concretes. Changes in shrinkage strains of OPC concretes for initial few weeks of curing have not been as uniform with time and with increasing content of air as BFS cement concretes. The highest increase in strains for increased air content was observed for OPC concrete with 8.1% of air. A less evident relationship has been identified for smaller content of entrained air. With total amount of air 3.4% i.e. for the lowest content of air voids – lower than 1.5%, the influence of air entrainment on strains was insignificant relative to nAE concrete for the first 28 days of curing. Increase in final strain of concrete with content of total air up to 8.1% in comparison with the strain of nAE amounted to about 220 * 10^{-6} and exceeded 40%.

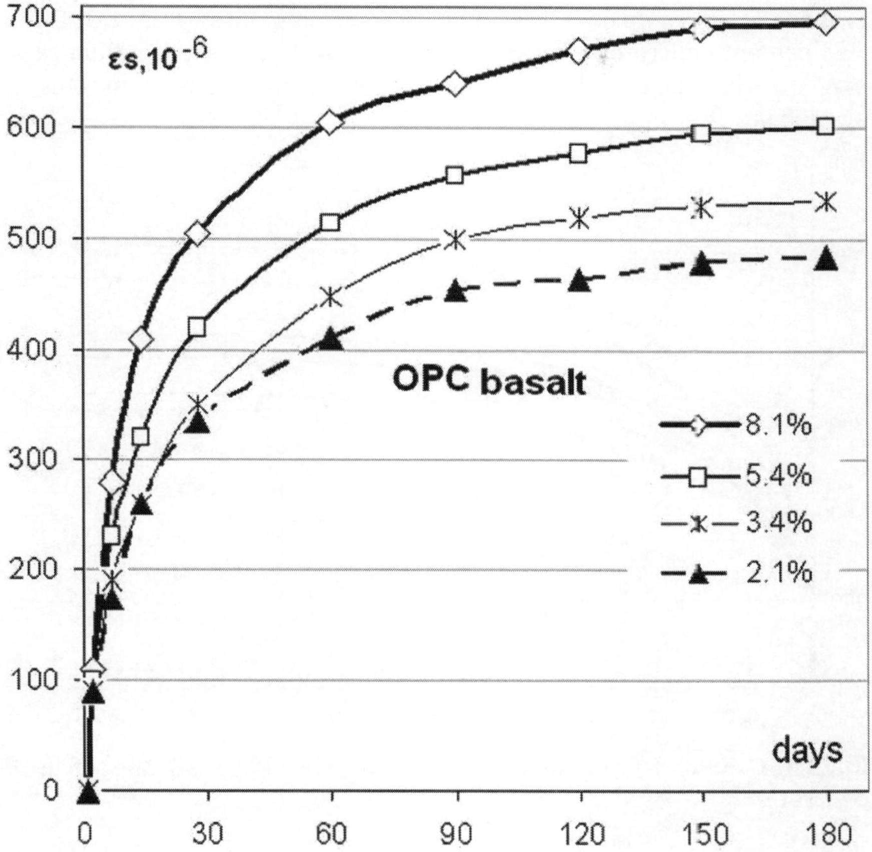

Figure 5. Influence of air content on shrinkage strains of OPC cement concretes with basalt aggregate.

3.3. Effect of cement type on shrinkage in AE concretes

Figure 6 shows diagrams of shrinkage strains of nAE concretes manufactured from gravel and carbonate aggregate with time. From the 14th day of the experiment, a visibly increasing difference occurred between the shrinkage strains of OPC concretes from and strains of BFS cement concretes. From the age of 30 days throughout the whole period of the experiment, there have been observed larger strains of OPC concretes in comparison to strains of concretes from cements containing BFS. The difference at the age of 300 days amounted to between 40% and 50% and from 25% to 40% for gravel concretes and carbonate aggregate ones, respectively. Analysing the strains of OPC and BFS cement concretes (Figure 6), it can be stated that the influence of slag on shrinkage in gravel and carbonate aggregate concretes is significant.

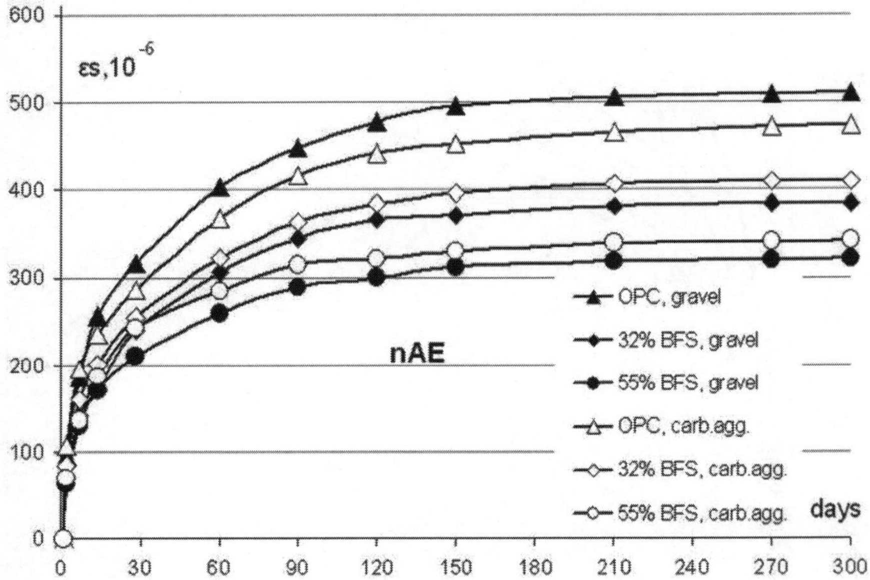

Figure 6. Influence of BFS cements on shrinkage strains of nAE concretes.

Figure 7 shows curves of shrinkage of AE gravel concretes from OPC and two cements containing 32% and 55% BFS. The beneficial influence of slag causing decrease in shrinkage was demonstrated. The differences between strains of AE concretes were significant. The lowest strains of concrete made with cement containing 55% BFS were determined.

Figure 8 shows shrinkage strains of AE concretes made from carbonate aggregate. Shrinkage strain of AE concretes was the highest when plain OPC was used. The increase in BFS content in blended cements decreased shrinkage strains of AE concretes. So the results suggest that cement of high content BFS is appropriate for low shrinkage concrete mixes which will be air-entrained.

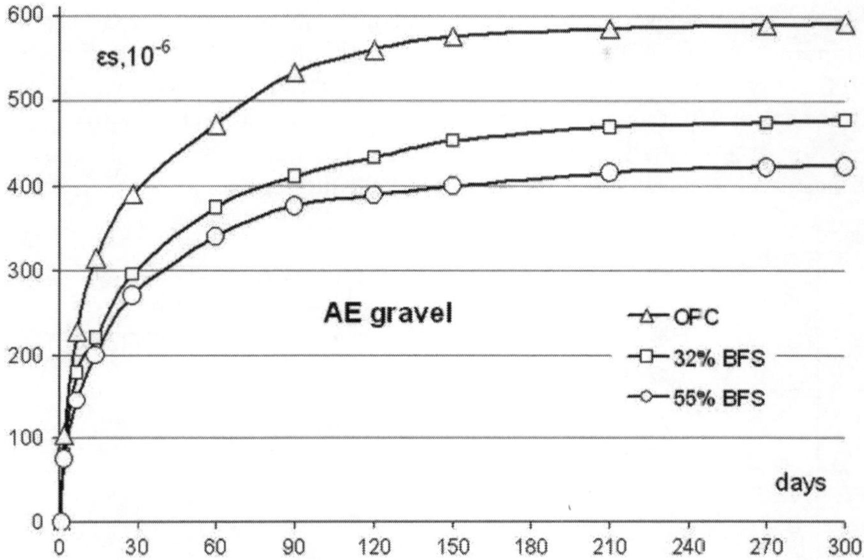

Figure 7. Influence of BFS content in cement on shrinkage strains of AE gravel concretes.

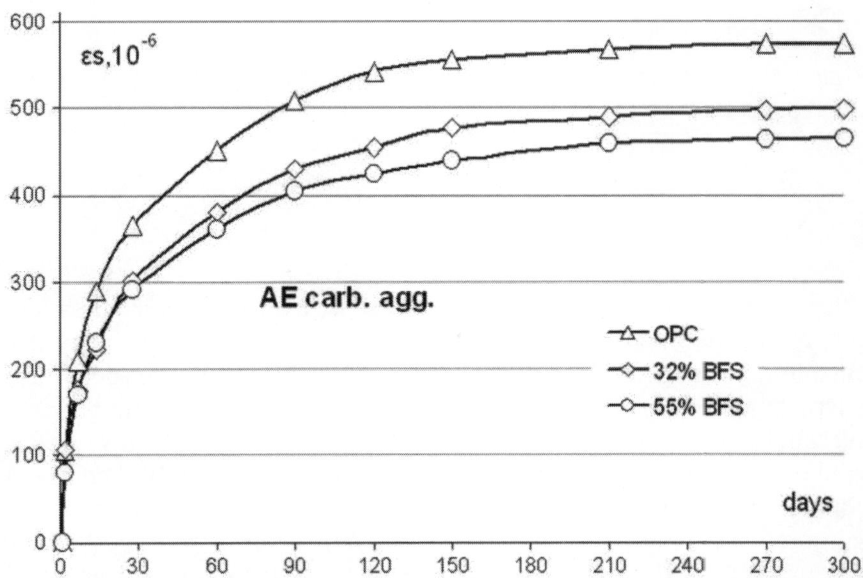

Figure 8. Influence of BFS content in cement on shrinkage strains of AE concretes from carbonate aggregate.

Figure 9 shows graphs of shrinkage strains of nAE and AE concretes manufactured from cements containing 14% LP compared to strains of OPC concretes. The shrinkage of limestone cement concrete was lower than that of OPC concrete regardless of the air entrainment (Figure 9). It has been confirmed that replacing 14% of cement with LP decreases concrete strains, when air entrainment was applied.

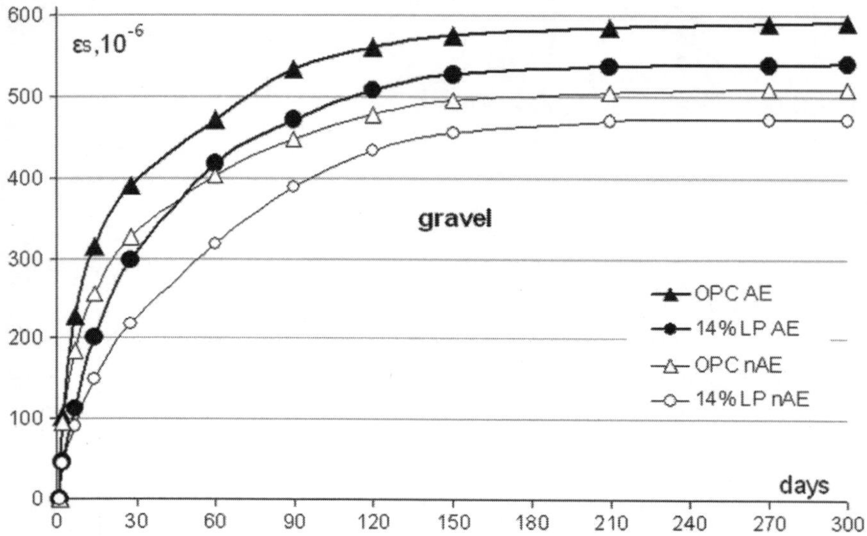

Figure 9. Influence of limestone cement on shrinkage strains of AE and nAE concretes.

3.4. Microstructure of AE concrete

Numerous SEM observation of microstructure of air-entrained concretes were carried out in order to find any answer why air entrainment increased shrinkage strains of concretes. In particular, the microstructure of concretes was observed in the range of cement paste surrounding air void shells. The SEM image (Figure 10a) shows an air void together with its shell as well as surrounding paste in OPC concrete containing 8.1% of air. The image (Figure 10a) shows that the microstructure of air void shell is dense and has no pores. Microstructure of the shell appears to be continuous and not as usual uniform for a component of hardened cement paste.

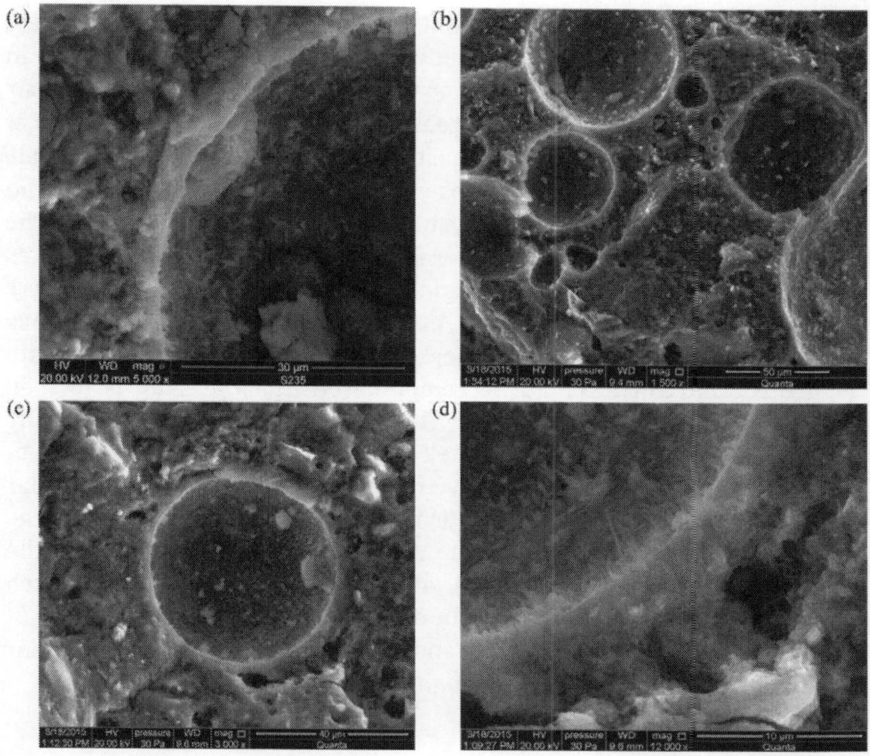

Figure 10. SEM images of air void shells and porous transition zones between the shell and bulk paste (a) air void and paste adjacent to the shell in OPC concrete of air content 8.1%; (b) arrangement of air voids in 50% BFS cement concrete of air content 7.8%; (c) overview of air void in 50% BFS cement concrete; (d) air void-cement paste interfacial transition zone in 50% BFS cement concrete.

The shell in aged OPC paste is made of CSH phase and the ratio of calcium-to-silicon of the air void shell is usually equal to 1.1 [19]. That is why the shell seems to stand out against a background of porous cement paste (Figure 10a).

It appears that between the paste and the shell there exists a certain discontinuity. On the inner surface of the air void shell, small ettringite crystals could be most likely observed. Around the shells, there appeared paste with porous and non-uniform microstructure creating the air void–paste interfacial transition zone. However, portlandite crystals have not been identified in any of the air void–paste interfaces observed in SEM images.

An example of pore arrangement in 50% BFS cement concrete containing 7.8% of air has been shown in Figure 10(b). All air voids in the Figure 10(b) are fine and smaller than 300 μm in diameter. Some air voids are close to each other. Whereas Figure 10(c) and (d) present SEM images of a typical air void–paste interface and air void shell. Microstructure of the shell is dense, continuous and uniform with no pores. It seems that very small crystals of ettringite are growing on the inner surface of the shell. In general, OPC and BFS cement concretes have similar shells. However, as compared to OPC paste (Figure 10a), there does not appear a visible discontinuity or distinct difference between shell phase and other phases in BFS cement paste (Figure 10c and d). It should be remembered that the shell is formed with low calcium phase CSH [19]. The low calcium phase CSH usually appears in hardened cement pastes containing substantial amount of slag. In air void interface there have been observed numerous irregular pores indicating its high porosity (Figure 10c and d). The width of the interface is around 30 μm from the void boundary [12]. As seen in the overall arrangement (Figure 10b), air voids may occur close to each other at high content of air voids in concrete. Due to the large width of the interfaces they may overlap and be interconnected, in particular when a content of the air voids is high.

4. STATISTICAL ANALYSIS AND DISCUSSION

4.1. Effect of air entrainment on shrinkage of concretes

Applying two-way analysis of variance with replication, a significant influence of air entrainment and cement type on shrinkage strains of gravel concretes at the age of 300 days has been demonstrated. The analysis results are presented in Figure 11. It is vital though that tests of simultaneous comparisons (post hoc analyses) have shown that differences between average results of shrinkage strains of AE and nAE concretes were statistically significant for all blended cements (the bars do not overlap).

The apparent influence of air entrainment on shrinkage behaviour of concretes may be related to an increase in volume of the cement paste despite a decrease in content of every component applied in 1 m^3 of the mix. The air voids are placed in cement paste, whereas the percentage of air entrainment is related to the whole concrete mixture volume. For instance, BFS cement concretes containing 3.7%, 5.2% and 7.8% air

(Table 7) as compared to nAE concrete containing on average 2% experienced an increase in volume of cement paste by respectively 17 dm^3/m^3, 32 dm^3/m^3 and 58 dm^3/m^3 as a result of air entrainment. Therefore, air comprised 11.6%, 15.8% and 22.5% of paste volume in AE BFS cement concrete mixes and 6.5% of paste volume in nAE concrete mix (Table 7). That is why the factor may not be irrelevant for the concrete strain as it is for its strength, which decreases with an increase in air content. According to proportionality of shrinkage to cement paste volume in nAE ordinary and high performance concrete mixes [16], the growth of paste volume should increase the overall shrinkage. Research on AE concretes cured for 180 days have shown increase in shrinkage strains with increase in air voids content. As compared to nAE concrete the increase in shrinkage strains of AE concrete as a result of air entrainment (Figure 4) was equal about 20%, 40% and 70% when the overall air content was 3.7%, 5.2% and 7.8% in AE concretes. This in turn means that shrinkage strain was consistent with air content. Therefore, it can be assumed that an increase in paste volume and decrease in aggregate due to air entrainment may have been the reason for larger shrinkage strains of air-entrained concretes.

Nevertheless, the difference between nAE concrete and AE concretes strains at the highest air contents is relatively high. Such a large difference suggests there could appear other reasons for higher strains of AE concretes. They may be related to higher porosity of interfacial transition zones around air voids, as shown in Figure 10(c) and (d), in particular that the air void–paste interfaces occupy a considerable fraction of the cement paste volume. As seen in Figure 10(b), their overlapping and interconnection are possible because the width of interface is relatively large at around 30 μm. High content of entrained air voids increases gaseous diffusivity even 2–3 times [12]. However, high vapour diffusivity favours quick moisture loss while drying of unsaturated concrete [15]. This in turn results in high rate of drying shrinkage. Therefore, the identified higher shrinkage strains of air-entrained concretes and their growth with increasing air content may be most likely related to more porous microstructure and bigger diffusivity of air void–cement paste interfacial transition zone, as it has been proved consistent with proportionality rule of shrinkage to paste content in nAE concrete mix.

Despite the increase in cement paste volume in concrete mixes, air voids entrained with AEA decreased the content of cement and all other components approximately by air entrainment percentage. Thus, the aggregate/cement ratio of AE and nAE concretes was the same.

Therefore, it may be assumed that this decrease in the content of cement did not affect shrinkage strains.

4.2. Effect of mineral additives content on shrinkage of AE concretes

In order to determine the relationship between test results of shrinkage of AE concretes and content of mineral additives replacing part of cement, in Figure 12 there have been marked with the points shrinkage strains of the 14 concretes tested cured for 300 days.

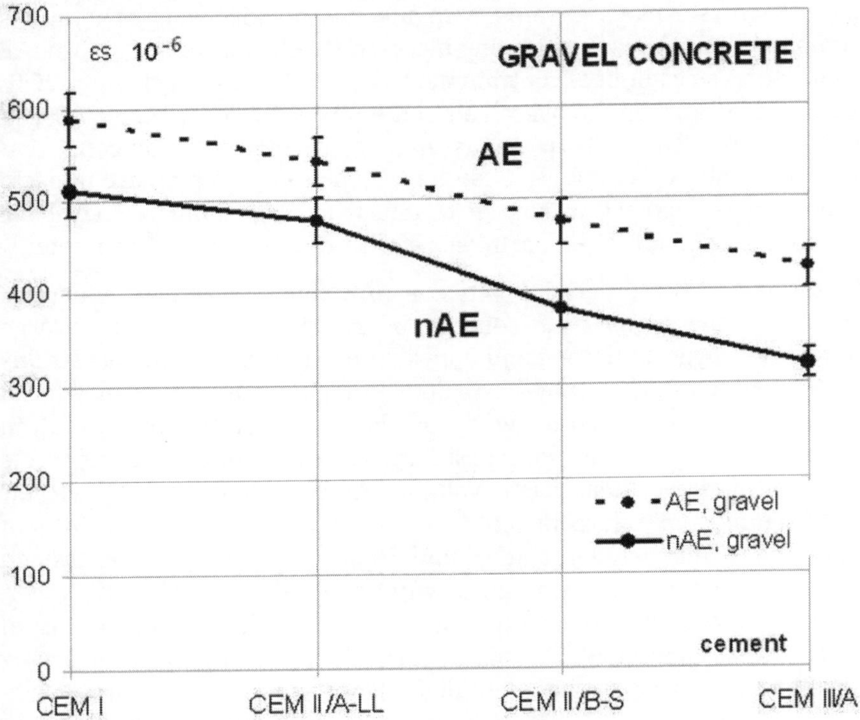

Figure 11. The effect of cement type and air entrainment according to two-way analysis of variance with replication on gravel concretes shrinkage at the age of 300 days. Vertical bars denote 0.95 confidence intervals.

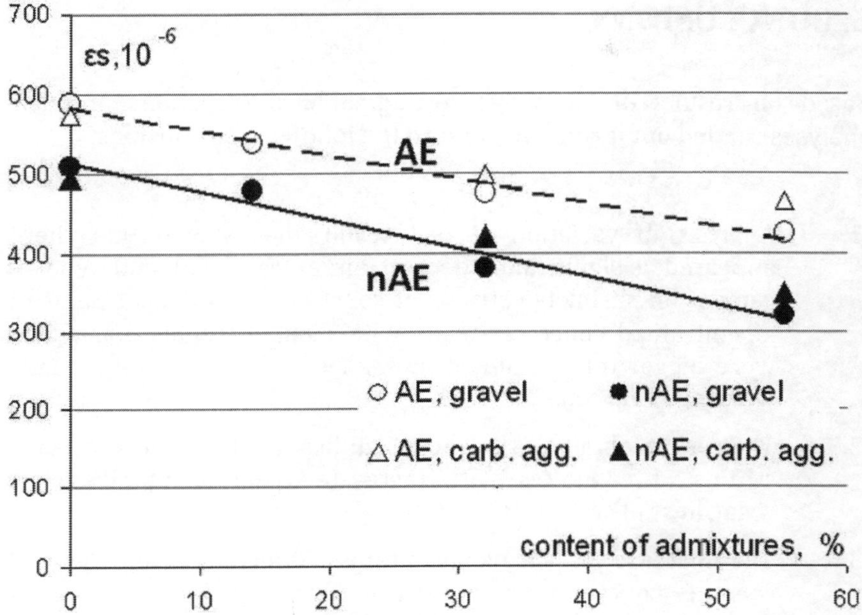

Figure 12. Relationship between shrinkage strains of concretes at the age of 300 days and content of mineral additives in cement (summative content of additives in the cement was marked on the axis of abscissae).

The cements used for the concretes contained LP and BFS in the amount as presented inTable 2. Determined after 300 days strains (Figure 12) decrease with mineral additives content regardless of type of additive or type of aggregate. The curve, which depicts average shrinkage strains of AE concretes depending on summative content of mineral additives in cements decreases by about 165×10^{-6}. However, the curve, which depicts strain results of nAE concretes decreases by about 190×10^{-6}. The difference is insignificant. Thus, the positive effect of the additives on reduction of shrinkage also appears when concrete is air-entrained.

Decrease in shrinkage of AE concretes together with the increase in content of LP or BFS can be explained with the decrease in amount of cement clinker whose relatively rapid hydration and drying is one of main sources of shrinkage. Besides, decrease in blended cement hydration kinetics, related to smaller reactivity of mineral additives results in decrease in the shrinkage strain. Due to very slow hydration of the slag its grains act as a microfiller which retards and decreases shrinkage of hardening cement paste as limestone powder does.

5. CONCLUSIONS

Based on results of shrinkage testing, SEM observations and the analyses carried out it could be noticed the following conclusions:

1. Analysis of variation of test results has shown that there appeared a significant effect of air entrainment and type of cement on shrinkage strains of concretes. Shrinkage strains of air-entrained concretes made with blended cements and OPC were bigger than strains of non-air-entrained concretes made with the same cements.

2. The results obtained showed a significant increase in shrinkage strains of concretes with increasing content of air voids, regardless of cement type.

3. The increase in shrinkage strains of air-entrained blended cements concretes with increasing air content may be most probably related to porous microstructure of air void–cement paste interfacial transition zone and overlapping and interconnection of the porous interfaces. The complex porous microstructure of cement paste adjacent to air void and in particular, the interconnection may increase vapour diffusion in AE concretes accelerating the moisture loss and drying shrinkage. Another reason for shrinkage increase may be increase in the volume of cement paste and decrease in the volume of aggregate in AE concrete.

4. Shrinkage of AE concretes just like nAE concretes was lowered with content of mineral additives used in blended cements. The decrease in shrinkage with the increase in the content of additives replacing cement can be explained with the decreasing amount of cement clinker which is the main source of shrinkage. Moreover, grains of slag and limestone powder act as microfillers which may retard and decrease shrinkage of hardening cement paste.

ACKNOWLEDGEMENTS

This study was supported by the Kielce University of Technology, within the research project "The Shrinkage of Air-Entrained Concretes".

REFERENCES

1. W. Kurdowski, The Chemistry of Cement and Concrete, Springer, Netherlands, 2014.
2. B. Craeye, G. De Schutter, B. Desmet, J. Vantomme, G. Heirman, L. Vandewalle, Ö. Cizer, S. Aggoun, E.H. Kadri, Effect of mineral filler type on autogenous shrinkage of self-compacting concrete, Cem. Concr. Res. 40 (2010) 908–913.
3. J.J. Thomas, H.M. Jennings, A colloidal interpretation of chemical aging of the CASAH gel and its effects on the properties of cement paste, Cem. Concr. Res. 36 (2006) 30–38.
4. P.C. Aitcin, The durability characteristics of high performance concrete: a review, Cem. Concr. Compos. 25 (2003) 409–420.
5. S. Tongaroonsri, S. Tangtermsirikul, Effect of mineral admixtures and curing periods on shrinkage and cracking age under restrained condition, Constr. Build. Mater. 23 (2009) 1050–1056.
6. E. Güneyisi, M. Gesoğlu, E. Özbay, Strength and drying shrinkage properties of self-compacting concretes incorporating multi-system blended mineral admixtures, Constr. Build. Mater. 24 (2010) 1878–1887.
7. L.M. Dellinghausen, A.L.G. Gastaldini, F.J. Vanzin, K.K. Veiga, Total shrinkage, oxygen permeability, and chloride ion penetration in concrete made with white Portland cement and blast-furnace slag, Constr. Build. Mater. 37 (2012) 652–659.
8. A. Itim, K. Ezziane, El-Hadj Kadri, Compressive strength and shrinkage of mortar containing various amounts of mineral additions, Constr. Build. Mater. 25 (2011) 3603–3609.

9. K.M. Lee, H.K. Lee, S.H. Lee, G.Y. Kim, Autogenous shrinkage of concrete containing granulated blast-furnace slag, Cem. Concr. Res. 36 (2006) 1279- 1285.

10. Y. Li, J. Bao, Y. Guo, The relationship between autogenous shrinkage and pore structure of cement paste with mineral admixtures, Constr. Build. Mater. 24 (2010) 1855–1860.

11. Z. Giergiczny, M.A. Glinicki, M. Sokołowski, M. Zielinski, Air void system and frost-salt scaling of concrete containing slag-blended cement, Constr. Build. Mater. 23 (2009) 2451–2456.

12. H.S. Wong, A.M. Pappas, R.W. Zimmerman, N.R. Buenfeld, Effect of entrained air voids on the microstructure and mass transport properties of concrete, Cem. Concr. Res. 41 (2011) 1067–1077.

13. D. Penev, M. Kawamura, Moisture diffusion in soil–cement mixtures and compacted lean concrete, Cem. Concr. Res. 21 (1991) 137–146.

14. S.F. Wong, T.H. Wee, S. Swaddiwudhipong, S.L. Lee, Study of water movement in concrete, Mag. Concr. Res. 53 (3) (2001) 205–220.

15. N. Jafarifar, K. Pilakoutas, T. Bennett, Moisture transport and drying shrinkage properties of steel–fibre-reinforced-concrete, Constr. Build. Mater. 73 (2014) 41–50.

16. A.M. Neville, The Properties of Concrete, fifth ed., Polski Cement, Cracow, 2012.

17. A.I. Rashed, R.B. Williamson, Microstructure of entrained air voids in concrete, Part I, J. Mater. Res. 6 (9) (1991) 2004–2012.

18. D.J. Corr, J. Lebourgeois, P.J.M. Monteiro, S.J. Bastacky, E.M. Gartner, Air void morphology in fresh cement pastes, Cem. Concr. Res. 32 (2002) 1025–1031.

19. M.T. Ley, R. Chancey, M.C.G. Juenger, K.J. Folliard, The physical and chemical characteristics of the shell of air-entrained bubbles in cement paste, Cem. Concr. Res. 39 (2009) 417–425.

20. G. Fagerlund, Influence of Pore Structure on Shrinkage, Strength and Elastic Moduli of Concrete. 44th Rep, Lund University of Technology, Lund, 1973.

21. P.W. Keene, The Effect of Air-Entrainment on the Shrinkage of Concrete Stored in Laboratory Air. TRA/331 Tech. Rep., Cem Concr Assoc, London, 1960.

7

CHAPTER

The Role of Lithium Compounds in Mitigating Alkali-Gravel Aggregate Reaction

J. Zapała-Sławeta, Z. Owsiak

Faculty of Civil Engineering and Architecture, Kielce University of Technology, Poland

HIGHLIGHTS

- Factors that affect the reactivity of polymineral glacial gravel aggregates were determined.
- The benefits of lithium nitrate to mitigate alkali-aggregate reaction were evaluated.
- The microstructure of the reaction products in mortars modified with lithium nitrate was studied.

ABSTRACT

Alkali reaction with reactive aggregates is an example of concrete internal corrosion. Cracks and damage to elements of a concrete structure are the result of the formation of an expansive alkali-silica gel. One of the ways to minimise the effects of this damaging reaction is to use lithium compounds as an admixture to concrete mixes. Lithium ions are believed to reduce the reaction between the alkalis and aggregate or minimise damaging expansivity by modifying chemical composition of the reaction products.

This paper presents the results from the tests carried out on mortars from reactive polymineral gravel with lithium nitrate according to the modified ASTM C1260 standard and ASTM C227. The findings show that lithium ions decrease mortar expansion until it reaches the levels markedly lower than those set out in the standards. A considerable reduction in the number of microcracks may indicate either lower swelling capacity of the gels or reduced reactivity of the aggregate, leading to the reduction in the quantity of damaging expansive products.

Keywords: Alkali-silica reaction; Lithium nitrate; Expansion; Gravel aggregate

1. INTRODUCTION

One of the causes of concrete deterioration is the reaction between siliceous aggregates and alkalis in the cement. Reactive aggregates contain reactive forms of silica which enter into reaction with sodium and potassium hydroxides quickly – opal, tridymite, cristobalite, acid volcanic glass – and those which react slowly – chalcedony, cryptocrystalline quartz and strained quartz [1]. Alkali reactivity of aggregates depends on their geological origin, mineralogical composition and texture. Identification of the factors that affect the gravel reactivity is difficult, and due to its polymineral character, qualitative and quantitative analyses of reactive constituents have to be performed to be able to choose proper methods for reducing reactivity potential. Gravel aggregate are commonly used in concrete, thus the minimisation of the alkali-aggregate reaction effects is critical [2].

The idea that lithium compounds can be used to mitigate negative effects of the aggregate reaction with sodium and potassium hydroxides first appeared in the 1950 s. Having conducted a comprehensive investigation on the inhibition potential of over 100 different compounds, McCoy and Caldwell reported lithium compounds to be most effective [3]. Worldwide interest in the lithium-bearing compounds has found lithium nitrate, used as a replacement for mineral admixtures, to have the highest effectiveness in preventing negative effects of alkali-aggregate reaction [4], [5], [6] and [7]. It has been demonstrated that this neutral and well soluble salt does not raise the pH value of the solution in the concrete pores, thus eliminating the risk of the pessimum effect [8]. Inappropriate doses of lithium ions in the form of lithium carbonate and lithium hydroxide lead to increased expansion and are deleterious to concrete, as observed by Diamond and Ong [9], Kawamura and Fuwa [10] during their studies of mortars containing LiOH and Li_2CO_3. The lithium compound introduced ions introduced into the concrete mix is thus of great importance. According to Feng et al., the effectiveness of lithium ions may vary for aggregates with different reactivity levels [11].

Several mechanisms of the reaction in the presence of lithium ions have been proposed. Some of them are based on increased stability of silica due to reduced pH of pore solutions or a change in their chemical composition [12]. It is presumed that amorphous or crystalline products containing silicon and lithium may form on the surface of reactive silica grains [13], acting as a protective barrier against Na^+ and K^+ ions. Other reaction mechanisms, not related to the Li^+ induced increase in silica stability, focus on the modification of the resulting gel products with lithium ions, rendering the modified gels less capable of swelling than traditional gels [13] and [14]. Crystalline lithium silicate may also form on the surface of the reactive silica. The literature reports the effect of Li^+ ions on the increased solubility of silica, which remains in the solution and prevents the gel from swelling [15].

2. MATERIALS AND METHODS

2.1. Aggregates

Polymineral glacial gravel aggregates, 0/16 mm, from the northern regions of Poland were used. Figure 1 shows the results of the X-ray analysis. The aggregates tested contain large amounts of quartz, calcite and plagioclase series with albite and anorthite, and lower amounts of clay minerals (illite, kaolinite) and dolomite.

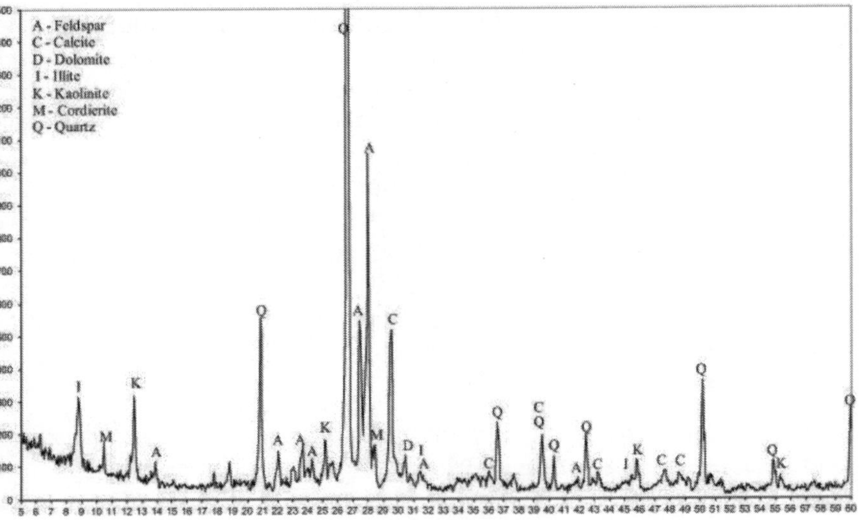

Figure 1. XRD diffraction pattern of gravel aggregate.

Mass loss results from the aggregate fraction measurements conducted according to PN-92/B-06714-46 [16] and the results for silica leachability according to ASTM C289[17]classify the gravel aggregate as potentially reactive (PN-92/B-06714-46) and reactive (ASTM C289) (Figure 2a, b).

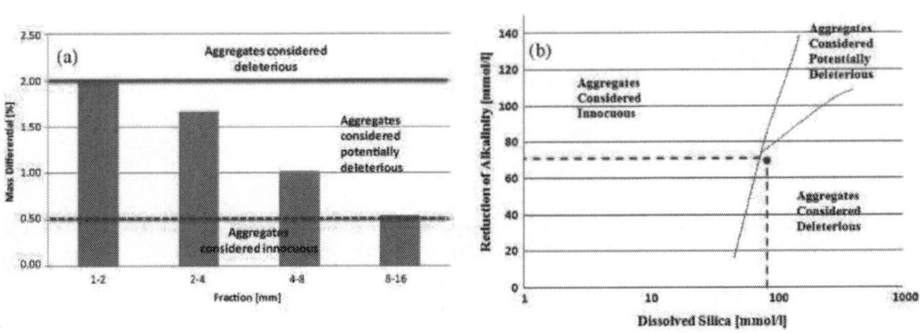

Figure 2. Reactivity Of Gravel Aggregate According To Specifications (a) PN-92/B-06714-46 (b) ASTM C289-94.

Those aggregate grain types, identified in macroscopic observations, which showed the greatest mass loss in the potential reactivity evaluation were subjected to a petrographic analysis. The following reactive components were found using an optical microscope: organodetritic sparite – micrite limestone, metamorphic quartz pyroxene shale with opal cement, quartz – glauconite sandstone with clay-carbonate binder including some chalcedony, and quartz grains in feldspar – biotite granite. Three main constituents, chalcedony, opal and strained quartz, are probably responsible for the reactivity of the aggregate. Table 1 shows the analysis of the quantitative composition of minerals in the aggregate.

Table 1. Reactive components in polymineral aggregate.

Rock type	Reactive constituent	Content [%vol.]
Quartz glauconitic sandstone	Chalcedony	2.1
Feldspathic granite with biotite	Quartz	15.7[a]
Quartz pyroxene shale	Opal cement	47.0
Organodetritic sparite – micrite limestone	Quartz	0.1

[a]Strained quartz.

2.2. Cement

Portland cement CEM I 42,5R containing 0.66% Na_2O_e was used. Table 2 shows the chemical composition of the cement.

Table 2. Cement chemical composition.

Material	SiO_2	Al_2O_3	Fe_2O_3	CaO	MgO	SO_3	K_2O	Na_2O	TiO_2	LOI	Cz. n.r.
Cement	20.20	4.80	3.00	61.70	1.80	2.70	0.78	0.15	0.45	3.00	1.00

LOI – loss on ignition, Cz.n.r. – HCl and Na_2CO_3 – insoluble parts.

2.3. Test methods

To evaluate the potential reactivity of the gravel aggregate, the accelerated ASTM C1260method was used together with the long-term ASTM C227 method [18] and [19]. The grain composition was prepared from the mechanically crushed gravel aggregate according to the requirements outlined in ASTM C1260 and ASTM C227. The bars with dimensions 25 × 25 × 250 mm were made with and without lithium nitrate. The lithium nitrate was added to the mix with the mixing water in the molar ratio of lithium ions to alkali ions [Li]/[Na + K] kept to 0.74 [11]. The bars were immersed in the solution with the [Li]/[Na + K] molar ratio as in the mortar bars (ASTM C1260) at 80 ± 2 °C and above the water table at 38 ± 2 °C (ASTM C227). The measurements were taken for 14 days in the accelerated method and at 14, 30, 60, 90, 180, 270 and 360 days in the long-term method. The microstructure of the resultant reaction products was studied under the scanning electron microscope with X-ray microanalysis.

3. TEST RESULTS

3.1. Mortar expansion test

Figure 3a, b and Table 3 present the results of the expansion tests for mortars modified with lithium nitrate and for non-modified mortars.

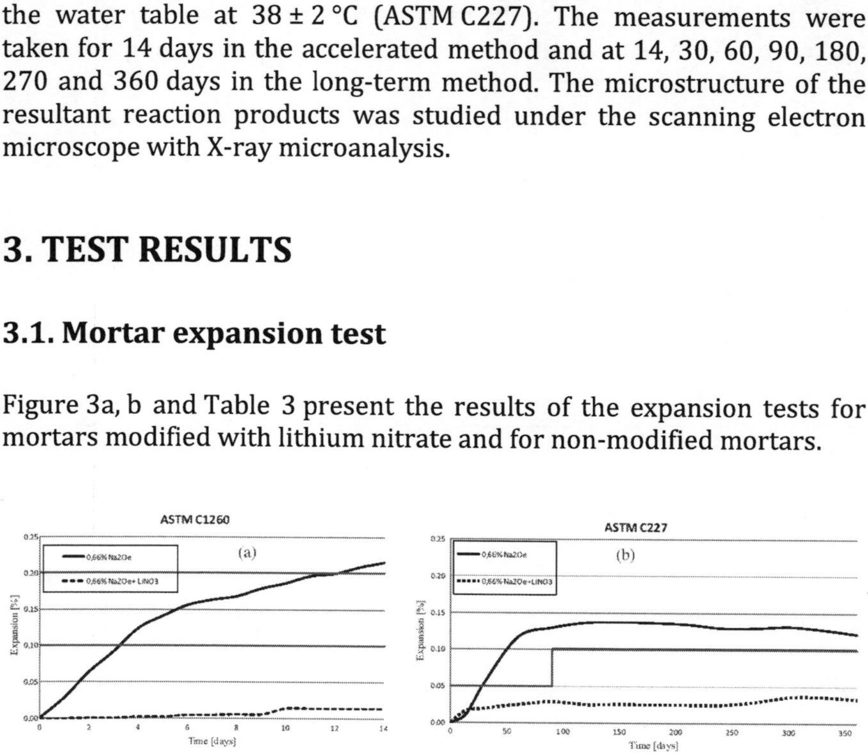

Figure 3. Expansion progress over time for mortars made with gravel aggregate, with and without lithium nitrate: (a) ASTM C 1260, (b) ASTM C227.

In the mortars without lithium nitrate, analysed using the short-term method, expansion exceeds the aggregate reactivity threshold of 0.1% after 14 days. High values of linear expansion were also recorded in the long-term method, which confirms the reactivity of the aggregate used. Higher linear expansion of the gravel aggregate mortars recorded in the

Table 3. Expansion of mortars with gravel aggregate at 14 days according to ASTM C1260 and at 360 days according to ASTM C227.

Method	Measurement time [days]	Mortar expansion [%]	
		Reference sample	Sample with $LiNO_3$
ASTM C1260	14	0.215	0.014
ASTM C227	360	0.122	0.033

accelerated method may result from different storage conditions. In the ASTM method C1260, the bars are subjected to high temperatures and external source of sodium hydroxide for a relatively short time. In the long-term method, the action of corrosive factors is "milder" owing to the lack of external alkali source, lower temperature and markedly longer exposure time.

In the mortars made with lithium nitrate admixture, elongation of the samples was limited to the safe level of up to 0.1% in the short-term method and 0.05% (up to 90 days) and 0.10% (after 90 days) in the long-term method. Expansion was reduced by about 93% in the mortars with gravel aggregate and lithium nitrate in the accelerated method, and by about 73% in the long-term method. To prevent washing the lithium ions out of the cement paste, the immersion solution was reinforced with lithium nitrate with the [Li]/[Na + K] molar ratio maintained at the same level as in the mortar bars. The differences in the percentage of expansion reduction in the two methods may result from the modification of the composition of the immersion solution.

3.2. Mortar microstructure

Microstructural observations of mortars without lithium compounds showed considerable quantities of microcracks in the samples tested using both methods (4a–b) (Figure 4). The gels formed in the particles contained large amounts of sodium and potassium.

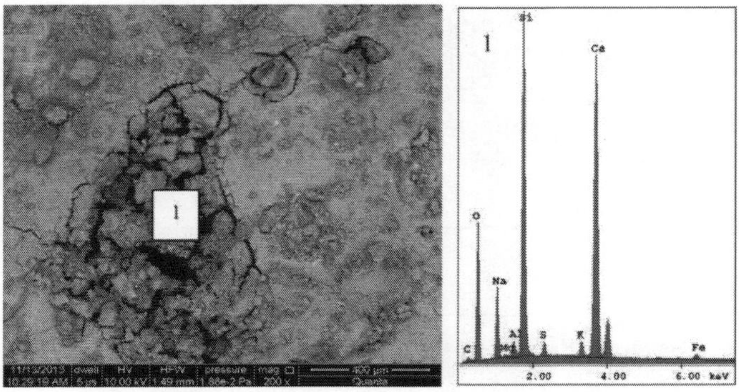

(a) (b)

Figure 4. Mortar samples without lithium nitrate, accelerated method, (a) microstructure and cracks, (b) X-ray microanalysis at point 1.

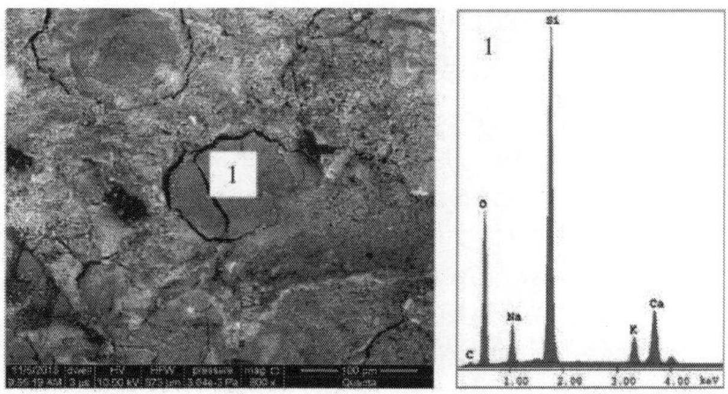

(a) (b)

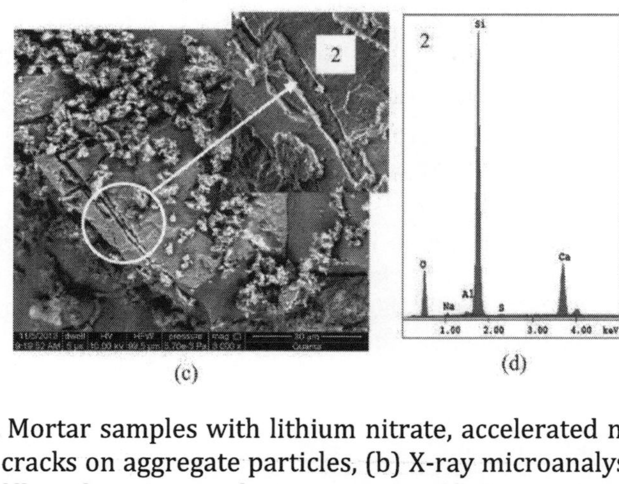

(c) (d)

Figure 5. Mortar samples with lithium nitrate, accelerated method, (a) resultant cracks on aggregate particles, (b) X-ray microanalysis at point 1, (c) gel filling the microcracks in aggregate, (d) X-ray microanalysis at point 2.

Figure 5 presents the findings from the observations of the microstructure of the mortars made with lithium nitrate. These mortars had much lower cracks in the region of aggregate particles and cement paste, compared with the control samples. The silicate hydrate gels showed lower contents of sodium at a variable content of calcium (Figure 5a, b). The products of the alkali-aggregate reaction, filling the microcracks in the aggregate, contained small amounts of sodium and potassium (Figure 5d).

4. CONCLUSIONS

1. The gravel aggregate used in this study reacted with the alkalis because expansion of the mortars exceeded the threshold of potential reactivity specified in ASTM C1260 and ASTM C227. Lithium compounds are effective inhibitors of expansion of mortars made with reactive gravel aggregates.
2. Products of aggregate reaction with sodium and potassium hydroxides filled the area of the aggregate, and the sodium-potassium-calcium silicate gel contained large amounts of sodium and potassium.
3. An addition of lithium nitrate in the molar ratio of Li/(Na + K) at the level of 0.74 helped decrease the expansion of mortars made with reactive gravel aggregate until it reached the safe, non-destructive level.
4. The differences between percentage reduction of expansion in the accelerated and long-term methods may result from the modification of the immersion solution composition. In the modified ASTM C1260 test method, due to the presence of lithium nitrate in the solution in which the mortar bars were stored, lithium leaching was limited and expansion decreased to 0.014%.
5. In the mortars with lithium nitrate, alkali content in the gel was small probably because the remaining lithium ions incorporated in the structure of the resultant products limit the sorption of sodium and potassium ions, thus contributing to the decrease of their swelling properties.

6. Further studies are necessary to better understand the mechanism of lithium compounds effect because available methods, for example, X-ray microanalysis, fail to analyse both the content and distribution of lithium due to its light weight.

REFERENCES

1. W. Kurdowski, The Chemistry of Cement and Concrete, The Association of Cement Producers, Krakow, 2010.
2. Z. Owsiak, J. Zapała-Sławeta, The lithium nitrate effect on the concrete expansion caused by alkali-silica reaction in concrete of gravel aggregate, Cem. Lime Concr. 82 (2015) 25–31.
3. W.J. McCoy, A.G. Caldwell, New approach in inhibiting alkali-aggregate expansion, ACI Mater. J. 22 (9) (1951) 693–706.
4. K. Era, T. Mihara, A. Kaneyoshi, T. Miyagawa, Controlling effect of lithium nitrite on alkali aggregate reaction, in: 13th International Conference on Alkali Aggregate Reaction in Concrete, Trondheim, Norway, 2008.
5. K. Kurtis, P. Monteiro, W. Meyer-Ilse, Examination of the effect of LiCl on ASR gel expansion, in: Proceedings of the 11th International Conference on AlkaliAggregate Reaction in Concrete, Quebec, 2000.
6. A. Leemann, L. Lörtscher, L. Bernard, G. Le Saout, B. Lothenbach, R.M. EspinosaMarzal, Mitigation of ASR by the use of LiNO3—characterization of the reaction products, Cem. Concr. Res. 59 (2014) 73–86.
7. X. Feng, M.D.A. Thomas, T.W. Bremner, K.J. Folliard, B. Fournier, Summary of research on the effect of LiNO3 on alkali–silica reaction in new concrete, Cem. Concr. Res. 40 (2010) 636–642.
8. T. Kim, J. Olek, Influence of lithium ions on the chemistry of pore solutions in pastes and mortars with inert aggregates, in: Proceedings of the 14th International Conference on Alkali-Aggregate Reaction, Austin, TX, 2012.
9. S. Diamond, S. Ong, The mechanisms of lithium effects on ASR, in: Proceedings of the 9th International Conference on Alkali-

Aggregate Reaction, Concrete Society of U.K., London, 1992, p. 269.

10. M. Kawamura, H. Fuwa, Effects of lithium salts on ASR gel composition and expansion of mortars, Cem. Concr. Res. 33 (2003) 913–919.

11. X. Feng, Effects and Mechanism of Lithium Nitrate on Controlling Alkali-Silica Reaction, University of New Brunswick, Canada, 2008.

12. B. Fournier, M.D. Thomas, D.B. Stokes, Influence of lithium products proposed for counteracting ASR on the chemistry of cement hydrates and pore solution, Cem. Concr. Res. 34 (2004) 1645–1660.

13. Z. Owsiak, J. Zapała-Sławeta, The course of the alkali-aggregate reaction in the presence of lithium nitrate, Ceram Silicaty 57 (2) (2013) 138–145.

14. X. Mo, B. Fournier, Investigation of structural properties associated with alkali–silica reaction by means of macro- and micro-structural analysis, Mater. Charact. 58 (2007) 179–189.

15. S. Diamond, Unique response of LiNO3 as an alkali silica reaction-preventive admixture, Cem. Concr. Res. 29 (8) (1999) 1271–1275.

16. PN-92/B-06714-46 Kruszywo mineralne – Badania – Oznaczenie potencjalnej reaktywnos´ci alkalicznej metodą szybka.

17. ASTM C 289-94, Potential Alkali-Silica Reactivity of Aggregates (Chemical Method), Annual Book of ASTM Standards.

18. ASTM C1260, Standard Test Method for Potential Reactivity of Aggregates (Mortar Bar Method), Annual Book of ASTM Standards.

19. ASTM C 227-10, Standard Test Method for Potential Alkali Reactivity of Cement – Aggregate Combinations (Mortar-Bar Method), Annual Book of ASTM Standards.

8

CHAPTER

Deterioration of Natural Hydraulic Lime Mortars, II: Effects of Chemically Accelerated Leaching on Physical and Mechanical Properties of Carbonated Materials

Phillip F.G. Banfill, Ewan M. Szadurski, Alan M. Forster

School of Energy, Geoscience, Infrastructure and Society, Heriot-Watt University, Edinburgh, Scotland EH14 4AS, United Kingdom

HIGHLIGHTS

- Reports accelerated leaching of carbonated natural hydraulic lime mortars.
- 1 M ammonium nitrate solution is a satisfactory leachant for these binders.
- Leaching removes binder, reduces strength and increases sorptivity.
- Leaching resistance increases in line with hydraulicity.
- Carbonated mortars are more resistant than uncarbonated ones.

ABSTRACT

Masonry using lime binders is very common in all regions of the world. Models for the future climate in northern regions predict conditions conducive for the materials used in mass masonry being saturated for longer periods and therefore at higher risk of binder leaching and consequent deterioration. Following work on uncarbonated lime mortars, ammonium nitrate leachant was used to accelerate the deterioration of carbonated mortars containing natural hydraulic lime binders. Leaching of binder significantly reduced the strength of the mortar and slightly increased its sorptivity. The resistance to deterioration increased in the order NHL2 < NHL3.5 < NHL5 binder, in line with the increasing hydraulicity of the binder. Carbonated mortars are more resistant to binder leaching than the corresponding uncarbonated materials.

Keywords: Natural hydraulic lime; Mortar; Binder; Dissolution; Leaching

1. INTRODUCTION

Traditional stone masonry, in the form of units bedded in lime mortar, is common worldwide. As a component of the external envelope of a building it must contribute to resisting the environment. There are three possible ways in which the interaction of water from the environment with masonry could change in response to climate change. First, in temperate northern zones the predicted increases in rainfall [1] and [2] suggest that masonry will remain wetter for longer, as a result both of wind-driven rain and of increased penetration into exposed infrastructural masonry (bridge piers, etc.). Second, where there is no damp-proof course at the base of a wall, evaporation from wall surfaces draws moisture from the ground and has the potential to drive significant amounts of water through the bed joints of masonry. Higher temperatures will increase the severity of this so-called evaporative pumping [3] and [4]. Third, the reduction of CO_2 emissions to mitigate climate change encourages retrofitting of thermal insulation to masonry buildings and this is likely to change the kinetics of moisture flow through masonry walls and potentially lead to build-up of moisture, especially if the insulation is of low water vapour permeability [5] and [6]. These higher moisture contents and water flows have the potential to leach binder from the mortar. In this process soluble components within the mortar dissolve and migrate to be re-deposited within pores, in construction voids or on the external face of masonry. This loss of binder has been reported by building professionals, with voids found in walls behind an apparently sound exterior [7] and [8], and is associated with progressive collapse in traditional buildings [9]. The first part of this study [10] reported, for the first time, the kinetics of lime binder leaching in uncarbonated natural hydraulic lime (NHL) mortars (such as those in the core of existing mass masonry or those in newly built work before atmospheric exposure has caused carbonation) and proposed a simple predictive model. This complementary paper aims to establish the rate of soluble binder leaching from carbonated natural hydraulic lime mortars (such as those expected to be near the surface and exposed to atmospheric CO_2) and its effect on properties and performance. The objectives were to determine the rate of calcium leaching in a range of carbonated NHL mortars, using ammonium nitrate as an accelerated leachant, and to assess the effect on strength and moisture handling characteristics. This would inform a method for predicting the service life of lime mortar.

Forster et al. [10] note that the lime binder components vulnerable to dissolution are portlandite (calcium hydroxide, $Ca(OH_2)$) and calcite (calcium carbonate, $CaCO_3$). The hydraulicity of the original binder influences a mortar's long term performance: less hydraulic limes have more $Ca(OH)_2$ and less of the hydraulic calcium silicate minerals and are therefore more reliant on carbonation for hardening, whilst more hydraulic limes have less $Ca(OH)_2$ and more calcium silicate. Consequently less hydraulic lime binders should be more susceptible to dissolution. This will also be aggravated in saturated, cold conditions because the solubility of both $Ca(OH)_2$ and $CaCO_3$ increases as water temperature decreases [11]. Accelerated leaching has been successfully used in cement and concrete studies [12] and [13] and Forster et al. [10] refined the method for lime binder to reflect its different composition and solubility, confirming that the loss of calcium from the hardened products changes the properties and performance of the uncarbonated mortar. The principle of chemical leaching is to promote reaction between calcium hydroxide or carbonate (in the binder material) and ammonium nitrate in the leachant solution, forming species that are highly soluble in water (see Eqs. (1) and (2)).

$$Ca(OH)_2 + 2NH_4NO_3 = Ca(NO_3)_2 + NH_3\uparrow + NH_4OH + H_2O \qquad (1)$$

$$CaCO_3 + 2NH_4NO_3 = Ca(NO_3)_2 + (NH_4)_2CO_3 \qquad (2)$$

Because the leachant concentration used in cement leaching proved too aggressive for NHL specimens, 1 M ammonium nitrate solution was used in both parts of this work[10,this paper] to accelerate the leaching of calcium (Ca^{2+}) ions.

2. MATERIALS, SPECIMEN MANUFACTURE AND CURING

2.1. Materials and characterisation

Mortars were prepared at a constant binder to aggregate ratio of 1:3 (by volume) using well graded silicaceous sand (Cloddach concrete sand, Elgin, Scotland). Using this sand ensured that binder leaching is not confused with dissolution of calcareous components from the aggregate. The particle size distribution was determined by sieving in accordance with BS EN 13139 [14], as previously reported [10]. The bulk density of the aggregate is 1500 kg/m^3. St Astier NHL2, NHL3.5 and NHL5, of bulk densities 550, 620 and 750 kg/m^3 respectively and complying with BS EN 459-1: 2010 [15], were used. The mineral compositions (X-ray diffractograms, Figure 1a–c) and manufacturer's anhydrous binder composition (Table 1) were the same as in part 1 of this work [10]. In these limes the amount of $Ca(OH)_2$ decreases as the hydraulicity of the material increases and this influences the potential for leaching [11].

Table 1. Chemical and physical properties of the NHL binders.

Binder	Bulk density kg/m^3	Mineralogical composition (%)							
		$CaCO_3$ unburnt	Insoluble	Free $Ca(OH)_2$	Compound				
					C_2S	C_3A	C_2AS	C_4AF	$CaSO_4$
NHL2	550	13	8	58	17	0.4	0.8	0.4	0.5
NHL3.5	620	25	9.6	25	35	0.5	1	0.5	0.8
NHL5	750	23	5.6	22	43	0.7	1.3	0.7	0.7

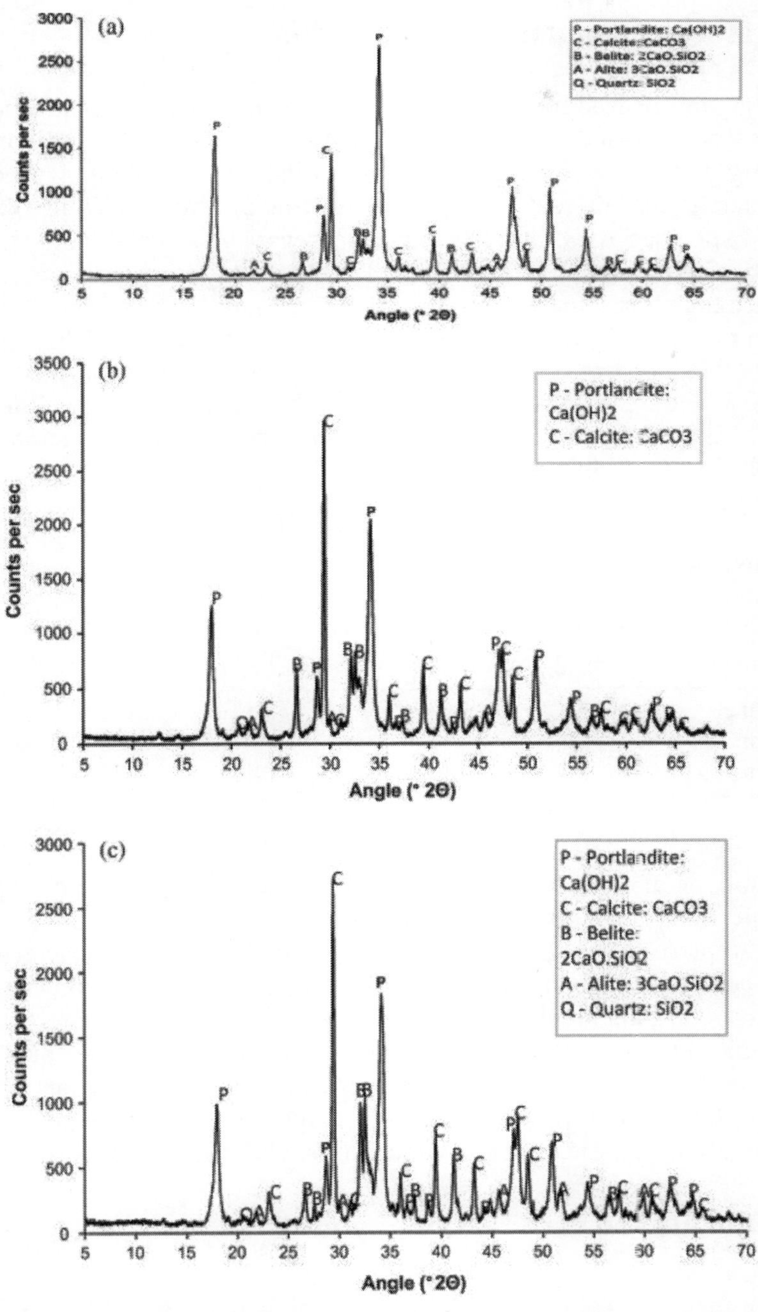

Figure 1. X-ray diffractograms for anhydrous Natural Hydraulic Lime (NHL) binders. (a) NHL2, (b) NHL3.5, (c) NHL5.

2.2. Mortar manufacture and batching

The mixing procedure for the manufacture of mortar test specimens was closely aligned to BS EN 459-2 [16] and BS EN 1015-2 [17] for the testing of lime mortars. The standards refer to the mixing procedure stipulated in BS EN 196-1 [18] for the testing of cements. Whilst BS EN 196-1 specifies a 1:3 ratio (1350 g of aggregate to 450 g of binder) with 225 g of de-ionised water, BS EN 459-2 varies the water content to ensure the mortar achieves a flow of 165 ± 2 mm for NHL2 and NHL3.5 and a flow of 185 mm ± 2 mm for NHL5 using a standard (BS EN 459-2) flow table apparatus [16]. In this programme, the proportions were further modified to integrate the relative bulk density for the different binder types to ensure a constant 1:3 ratio by volume was maintained for all mortars (Table 2).

Table 2. Mortar mix data.

Lime	Mass of material kg			Sand/lime by mass	Water/lime by mass	Flow value mm (±2 mm)
	NHL	Sand	Water			
NHL2	1.63	7.50	2.63	4.6	1.61	165
NHL3.5	1.89	7.50	2.73	4.0	1.44	165
NHL5	2.05	7.50	2.75	3.7	1.34	185

The mixing and casting of the mortars followed BS 459-2 (2010) [16], except that 160 × 40 × 40 mm lightweight polystyrene moulds were used in place of the specified steel gang moulds, in order to reduce the load on the shelving in the environmental curing cabinet. The polystyrene gang moulds were not treated with releasing agent to avoid possible adverse effects on the mould material or the mortar. Table 2 shows the quantities required for 8 L (0.008 m^3) of mortar. Fifteen 160 × 40 × 40 mm prisms were produced for each binder and carbonated before leaching, of which three were used as controls.

2.3. Curing and carbonation regime

The specimens were cured in a TAS series 3 environmental cabinet at 100% RH and 20 °C for 6 days, after which they were de-moulded and the RH level in the cabinet was reduced to 65% (±5%). This lower humidity promotes the carbonation reaction and, according to

Lawrence [19], the specimens are expected to be dry enough for carbonation to start by 14 days. An injection system maintained a constant CO_2 level of 800–1000 ppm to accelerate the carbonation process to achieve the target depth of 20 mm (marking full carbonation of the 40 × 40 mm prisms) by the end of 56 days, and the total curing and carbonation time chosen to be in line with Pavia and Toomey [20] and to accord with part 1 [10].

3. EXPERIMENTAL PROCEDURE

3.1. Accelerated leaching

The leaching procedure was the same as described in part 1 of this work [10]. After 56 days of curing, specimens contained in individual wire cages (1 M ammonium nitrate experiments) or placed on glass spacers (deionised water experiments) were immersed for up to 169 days at 20 ± 2 °C in a stirred tank. The control specimens were immersed in de-ionised water in parallel treatments for the initial 36 days. In each case up to 20 specimens were exposed to 15 L of leachant.

3.2. Physical and chemical testing

Specimens were removed from solution after 1, 4, 9, 16, 25, 36, 100 and 169 days and tested as follows. Unfortunately, insufficient specimens were available to complete all tests at 100 and 169 days, and it was necessary to curtail the water leaching after 36 days, as noted below.

3.2.1. Flexural Strength

Flexural strength was tested according to BS EN 459-2 [16] using three point bending in a Lloyds universal MK5 instrument of 5 kN maximum load capacity. Force was increased until failure occurred and the specimen broke in two. The two halves were preserved for other tests.

3.2.2. Depth of carbonation

The fractured surfaces were sprayed with 1.0% phenolphthalein solution to give an indication of the penetration of the carbonation front as a result of the curing and accelerated carbonation process.

3.2.3. Sorptivity

The specimens were oven dried at 85 °C for 24 h and then their sorptivity determined [3] and [10]. Each specimen was allowed to cool and then five faces were coated with resin, leaving the 40 × 40 mm fracture surface uncoated. Each specimen was supported on a pair of glass rods located in the bottom of a container and the uncoated face immersed to a depth of 1 mm in de-ionised water. Measurements of mass increase were taken at intervals over about one hour. The sorptivity determined from the slope of the straight line relationship between mass and $t^{1/2}$, is expressed in units of mm/min$^{1/2}$[3]. It may be noted that sorptivity is equivalent, in principle, to the water absorption coefficient, determined according to BS EN ISO 15148 [21]. A sorptivity of 1 mm/min$^{1/2}$ is equal to a water absorption coefficient of 0.13 kg/m^{2}s$^{1/2}$.

3.2.4. Compressive testing

Compressive strength was tested in accordance with BS EN 459-2 [16] using the remaining half of each prism. The fractured half was trimmed with a diamond saw (dry blade) to produce a 40 mm cube which was then tested in the Lloyds universal MK5 testing machine.

3.2.5. Petrographic analysis

Thin sections, prepared and impregnated with blue dyed resin to highlight porosity, were investigated by petrographic microscopy and a quantitative assessment to identify binder, aggregate and porosity undertaken by the point counting method previously described[10].

4. RESULTS

Physical testing results for all carbonated specimens subjected to ammonium nitrate and water leaching are presented here.

4.1. Degree of carbonation

The absence of pink colouration on the fractured surfaces when sprayed with phenolphthalein solution confirmed that the procedure described in Section 2.3 resulted in fully carbonated specimens at the start of the leaching treatment although it is possible that there may be a small amount of residual $Ca(OH)_2$. Recognising that the proportion of free $Ca(OH)_2$ in the NHL binders varies [10] and carbonation converts the $Ca(OH)_2$ to $CaCO_3$, the resulting $CaCO_3$ content in the mortars decreases in the order NHL2 > NHL3.5 > NHL5.

4.2. Compressive and flexural strength results

Table 3 shows that there is a clear difference between the specimens leached in ammonium nitrate and in de-ionised water. In water, carbonated specimens of NHL2, i.e. samples that have a high proportion of $CaCO_3$, lost strength, whilst those of NHL5, with more reliance on hydraulic set, gained strength over the test period. This is probably associated with the favourable curing environment for the development of the products of hydration, especially C–S–H.

Table 3. Compressive, flexural strength and sorptivity results (each entry denotes a single test on a single specimen).

Binder	Leachant	Compressive strength MPa				Flexural strength MPa				Sorptivity $mm/min^{1/2}$		Change in sorptivity %
		0 d	36 d	100 d	169 d	0 d	36 d	100 d	169 d	0 d	36 d	
NHL2	Water	1.14	0.94			0.31	0			1.96	1.90	−3.1
NHL3.5	Water	2.69	2.07			0.7	0.95			1.55	1.77	+14.2
NHL5	Water	2.48	3.02			0.74	1.83			2.11	1.08	−48.8
NHL2	NH_4NO_3	1.14	0.75	0.25	0.1	0.31	0.18	0.32	0.21	1.96	1.98	+1.0
NHL3.5	NH_4NO_3	2.69	0.91	0.16	0.5	0.7	0.38	0.29	0.21	1.55	1.91	+23.2
NHL5	NH_4NO_3	2.48	1.3	1.09	0.68	0.74	0.63	0.57	0.29	2.11	1.85	−12.3

In contrast, the effect of ammonium nitrate leaching on strength is significant, giving generally progressive reductions in compressive and flexural strength. Strength loss was more rapid with NHL2 than with the other binders and this is consistent with its higher proportion of $CaCO_3$[10], which reacts according to Eq. (2) above.

4.3. Sorptivity

Due to lack of specimens, sorptivity data is available only for 0 and 36 days leaching.Table 3 shows that 36 days water and ammonium nitrate leaching has no effect on sorptivity of carbonated NHL2, increases sorptivity of carbonated NHL3.5 and decreases sorptivity of carbonated NHL5. This behaviour is not consistent with the strength reductions observed in NHL2 and NHL5 in both leachants. Possibly in

NHL5 there is continuing hydration of C_2S during the leaching and the resulting formation of C–S–H fills the pores and counteracts any loss of $CaCO_3$. The significant increase in sorptivity observed in NHL3.5 could be because there is less C_2S and hence less C–S–H can form to offset the loss of $CaCO_3$.

4.4. Petrographic analysis

Figure 2, Figure 3 and Figure 4 show the effect of 36 days ammonium nitrate leaching on mortars, viewed in thin section. Blue is porosity, grey is aggregate and dark grey/brown is binder.

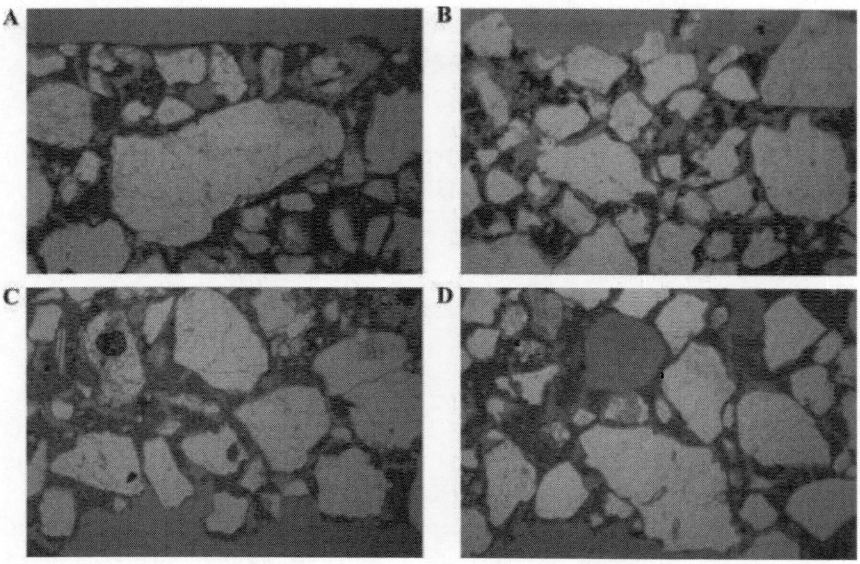

Figure 2. Carbonated NHL2 mortar thin sections: (A) initial outer, (B) 36 days outer, (C) initial inner, (D) 36 days inner.

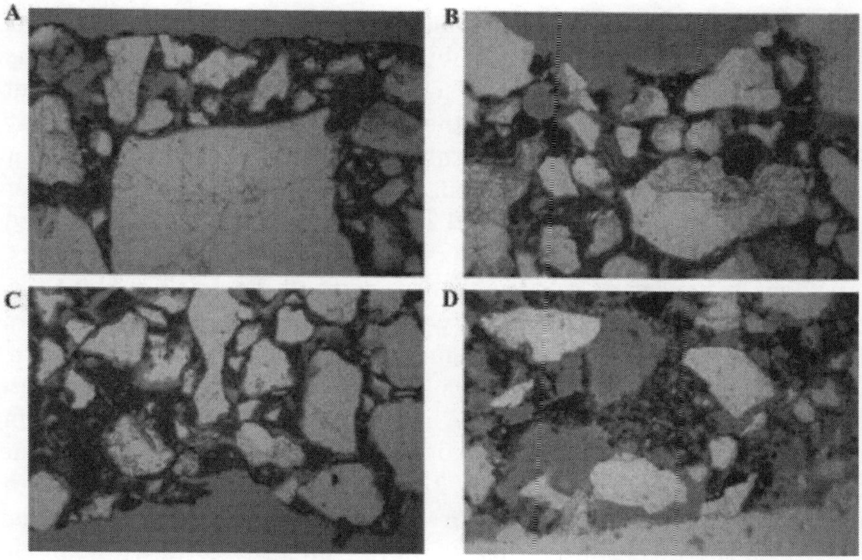

Figure 3. Carbonated NHL3.5 mortar thin sections: (A) initial outer, (B) 36 days outer, (C) initial inner, (D) 36 days inner.

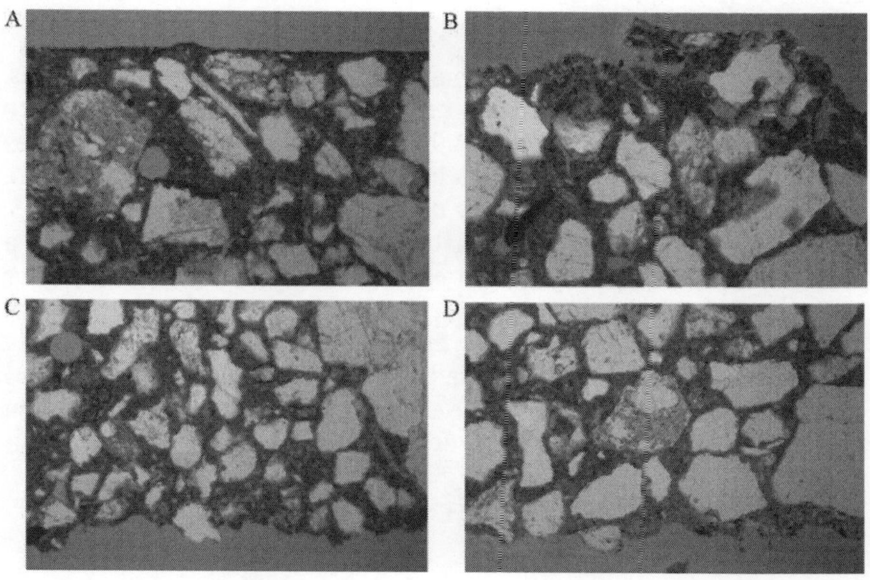

Figure 4. Carbonated NHL5 mortar thin sections: (A) initial outer, (B) 36 days outer, (C) initial inner, (D) 36 days inner.

The thin sections show features typical of carbonated NHL mortars. The aggregate is well graded and distributed and the bond (evaluated as the extent of contact between aggregate and binder) is good in NHL5 but less so in NHL2 and NHL3.5, with the latter in particular showing locally poor bonding. The bonding is slightly impaired towards the inner regions. Cracks increase in abundance and width towards the inner regions and are wider in the NHL2 and 3.5 mortars but smaller and better defined in the NHL5. This is because with the stronger binders, cracking during curing is more likely to occur at the weakest point (i.e. the aggregate-binder interface) whereas with the weaker binders cracking occurs preferentially within the binder. Carbonation is diffuse, with the typical crystalline components too small to be observed with the petrographic microscope. When the mortars are leached in ammonium nitrate, moderate binder depletion is evident throughout the NHL2 and 3.5 samples (Figure 2 and Figure 3, images B and D) with widening of microcracks, minor loss of bond and enlargement of voids. In NHL3.5 there is complete loss of mortar integrity towards the inner region. NHL5 (Figure 4, images B and D) shows only minor to moderate depletion with its overall integrity maintained but compromised at the outer surface. Compared to the uncarbonated mortars [10] the loss of binder is less significant in every case confirming that carbonated mortar is more resistant to leaching.

These trends are supported numerically by point counting analysis, which was done on different thin sections. This identifies the presence or absence of binder at each of 600 points across the section but does not take into account any changes in binder density and the results are conveniently presented in ternary diagrams [22]. As shown in Figure 5, leaching of binder would be expected to move the composition of a mortar parallel to the binder–porosity axis in the direction of higher porosity. Figure 5 also shows the location of the zone enlarged in Figure 6 and Figure 7. Figure 6 shows the effect of 36 days leaching in ammonium nitrate on carbonated mortars. For comparison, Figure 7 shows the effect of 36 days leaching in ammonium nitrate on uncarbonated mortars, using the results presented in part 1 [10].

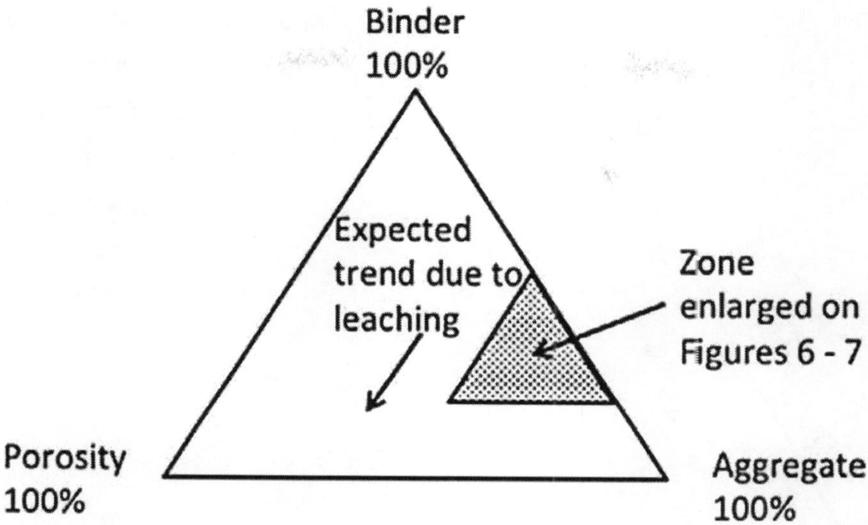

Figure 5. Schematic ternary diagram for mortar thin section point counting data.

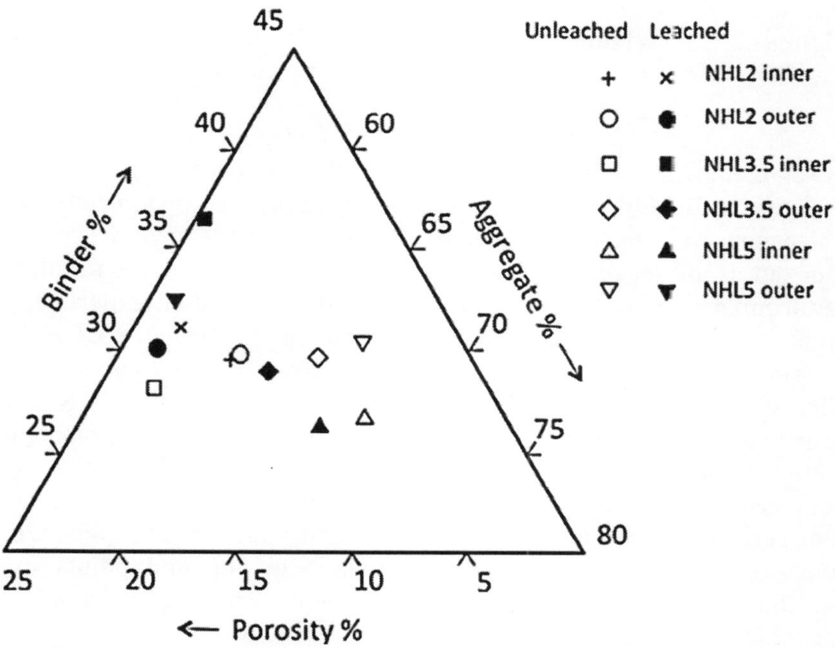

Figure 6. Thin section point counting data for carbonated mortars showing the effect of leaching in ammonium nitrate for 35 days.

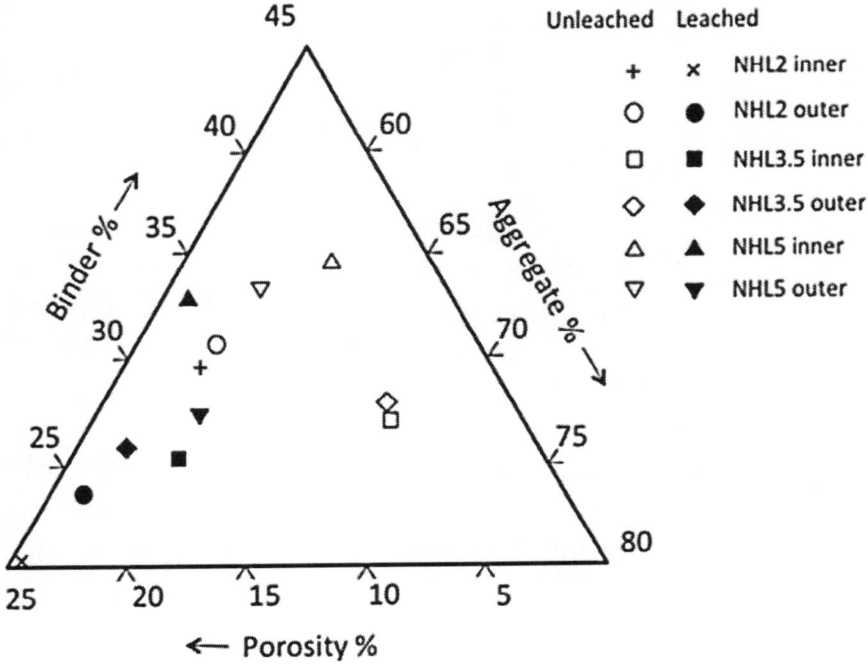

Figure 7. Thin section point counting data for uncarbonated mortars showing the effect of leaching in ammonium nitrate for 36 days.

Recognising that there is a 4% experimental error on the points in Figure 6 and Figure 7[10], the differences in composition between unleached and leached specimens are within experimental error for four out of the six carbonated mortars, whilst the differences for all the uncarbonated mortars are both significantly larger than experimental error and tend to align with the vector predicted from Figure 5. Even so, the small differences for the carbonated mortars (Figure 6) also tend to align with the predicted vector. The thin section for the NHL3.5 inner zone shows the reverse of the predicted vector: a lower porosity/higher binder content in the leached than in the unleached specimen. This is not consistent with the image shown in Figure 3D, which is of a different thin section taken from the same zone and shows much higher and coarser porosity. This suggests incomplete mixing and points to a possible reason for the observed anomalous leaching resistance of carbonated NHL3.5 mortar, discussed further below.

5. DISCUSSION

All the results presented here relate to carbonated natural hydraulic lime mortars: they can be taken as being relevant to mortar near the surface of masonry constructions, which has had the opportunity during its lifetime to dry sufficiently to allow carbonation to occur, i.e. it has not been continuously saturated, even though moisture may be present now or in the future to facilitate binder leaching. Strength measurements have established that ammonium nitrate successfully accelerates the leaching of binder from carbonated mortars, just as for uncarbonated mortars [10]. However, the reductions in strength are less marked than for uncarbonated mortars: whereas uncarbonated mortars lost 96–99.5% of their compressive strength over 169 days [10], the carbonated mortars lost 73–91% over the same period. Sorptivity of carbonated mortar is affected by leaching, and this agrees with the broad trend in porosity shown by the point counting analysis. This is also less severe than in the uncarbonated mortars [10], for which sorptivity increased by between 30% and 150%. These results suggest that carbonated mortars are more resistant to leaching than uncarbonated mortars, and this is in line with the relative solubilities of $CaCO_3$ and $Ca(OH)_2$, the predominant binder material in the respective mortars. Additionally, whereas in uncarbonated mortars the resistance to leaching is strongly influenced by the relative proportion of C–S–H within the matrix, with the most hydraulic lime (NHL5) showing the highest resistance, in carbonated mortars this effect is less dramatic as the lowering of binder solubility that comes with carbonation mitigates the influence of hydraulicity.

In part 1 [10] the leaching rate of uncarbonated mortars was quantified by the size of the alkaline zone visible in the fracture surface, when sprayed with phenolphthalein. This approach is not possible in these fully carbonated mortars because, by definition, there is no alkaline zone visible. Therefore another approach based on the decrease of strength will be adopted. The data presented in Table 3 above and in Table 3 of part 1 [10] suggest that the compressive strength decreases exponentially with time of leaching according to a relationship of the form:

$$fcu(t)/fcu(0)=e^{-bt} \quad (3)$$

where $f_{cu}(t)$ is the strength after leaching for time t and $f_{cu}(0)$ is the strength before leaching. A graph of $\ln(f_{cu}(t)/f_{cu}(0))$ against t should give a straight line of slope $-b$ and b is therefore a first order rate constant for the strength reduction process. Combining data for uncarbonated mortars from Table 3 of part 1 [10] and for carbonated mortars from Table 3 above, Figure 8 confirms that the logarithm of the strength, relative to that of the untreated specimens, decreases linearly with time, giving the values of b summarised in Table 4. For clarity, the regression lines are omitted from Figure 8 but it can be noted from Table 4 that, in four of the six sets of data, the value of R^2 (four points in each graph) is high. The value of b is consistently larger for uncarbonated than carbonated mortars made with all limes, and decreases in the order NHL2 > NHL5, reflecting the more rapid strength reductions in uncarbonated mortars and those of less hydraulic binders. However, the uncarbonated NHL3.5 mortars appear to be more resistant than both NHL2 and NHL5, whereas the carbonated NHL3.5 mortars are equally resistant to the NHL2 and less resistant than NHL5.

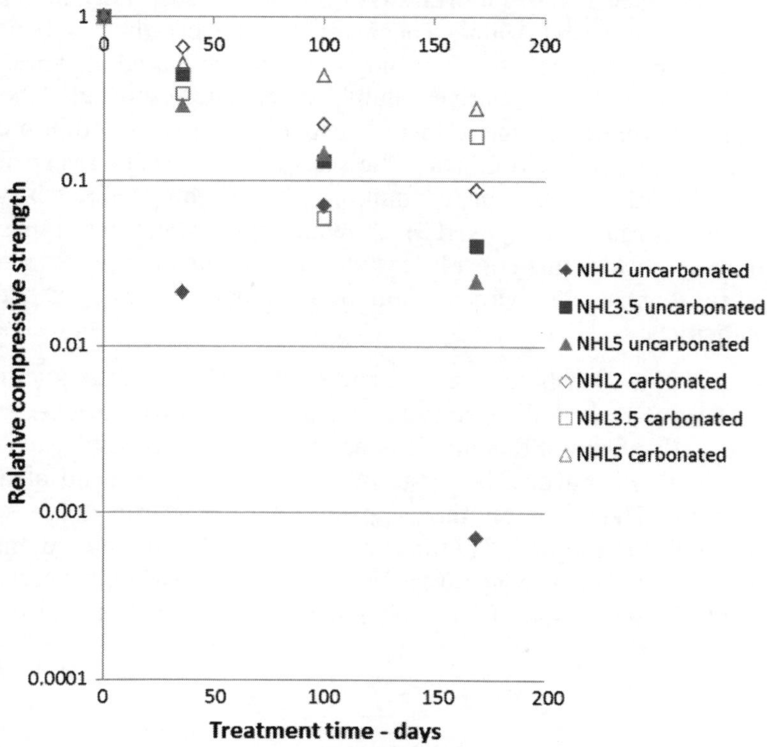

Figure 8. Strength reduction in uncarbonated and carbonated NHL mortars leached in ammonium nitrate.

Table 4. Strength reduction kinetics in NHL mortars treated with 1 M NH_4NO_3.

	NHL2 uncarbonated	NHL3.5 uncarbonated	NHL5 uncarbonated	NHL2 carbonated	NHL3.5 carbonated	NHL5 carbonated
Rate constant, b (day^{-1})	0.041	0.019	0.022	0.015	0.015	0.008
R^2	0.71	0.995	0.96	0.996	0.33	0.85
Estimated time to 20 mm leached depth (days)	34	73	63	92	92	173
k (Eq. (5)) (mm/day$^{1/2}$)	3.43	2.34	2.51	2.09	2.09	1.52
Leaching resistance ($1/k$)	0.29	0.43	0.40	0.48	0.48	0.66

Now, figure 12 of part 1 [10] shows that a leached depth of 20 mm is achieved in uncarbonated specimens in 42 days for NHL2, 50 days for NHL3.5 and 73 days for NHL5 mortars. The kinetic parameters in Table 4 show that these times correspond to residual strengths ($f_{cu}(t)/f_{cu}(0)$) of 0.18, 0.39 and 0.20 respectively, giving a mean value of 0.25 (i.e. residual strength is 25% of untreated strength). Assuming that the same relationship between strength and leached depth applies to carbonated mortar, where $CaCO_3$ is leached (Eq. (2)), and to uncarbonated mortar, where $Ca(OH)_2$ is leached (Eq. (1)), then the time to a residual strength of 25% gives an estimate of the time to a 20 mm leached depth. These estimated times are also shown in Table 4. Finally, part 1 [10] developed a relationship between leached depth h and treatment time t (Eq. (4)), where k is a constant, the reciprocal of which is an index of resistance to leaching (which is also shown in Table 4):

$$h = kt^{1/2} \tag{4}$$

Table 4 shows that the time to a leached depth of 20 mm in uncarbonated specimens as estimated by this procedure is slightly different from that directly measured by phenolphthalein spraying, given in figure 12 of part 1 [10]. However, for the sake of consistency in the analysis of the different materials, the values from Table 4 will be adopted here. Leaving aside the data for the NHL3.5 mortars, the index of resistance to leaching ($1/k$) is 38% higher for NHL5 mortar than for NHL2, whether uncarbonated or carbonated, and is 65% higher in carbonated than in uncarbonated mortar, whether NHL2 or NHL5. This is in agreement with the explanation, offered previously, that the more hydraulic lime is more resistant to leaching and that conversion of the $Ca(OH)_2$ to $CaCO_3$ by carbonation makes the mortar more resistant to leaching. It is particularly striking that the percentage increases in the index of resistance agree so closely. It may be noted that the square root dependence in Eq. (4) means that a 1.65-fold increase in the index of resistance as a result of carbonation increases the time to a given depth of leaching by a factor of 2.75.

Considering the results for NHL3.5 and the position of this binder in the sequence, the regression line for the strength decay of carbonated mortars (Figure 8) has an R^2 value of 0.33. This is not statistically significant and this implies a wide confidence interval on the value of b and hence on the estimated leaching resistance. This is probably related to the anomalous results of the thin section analysis (Figs. 3D and 6) discussed above. If the NHL3.5 mortars were incompletely mixed or compacted a wider scatter in the strength results would be expected, particularly as each point in Figure 8 is from a single specimen. A poorly compacted specimen would have low strength and if this were the single carbonated NHL3.5 specimen in Figure 8 which shows much lower relative strength after 100 days leaching it would have a substantial effect on the slope of the strength reduction regression line. This suggests that the true regression for NHL3.5 might lie between those for NHL5 (smaller strength reduction) and NHL2 (larger strength reduction), but this cannot be confirmed without repeating the experiment. Thus the data are not inconsistent with placing carbonated NHL3.5 mortars in order of leaching resistance between NHL2 and NHL5 mortars. Also, in uncarbonated mortars the values estimated from the strength decay are different from those obtained by direct observation of the carbonation front (figure 12 of part 1 [10]): the latter place uncarbonated NHL3.5 between NHL2 and NHL5 mortars. Thus, again there is evidence in support of a trend of increasing leaching resistance in the order NHL2 < NHL3.5 < NHL5. Therefore, it seems safe to say that high confidence can be ascribed to the values of leaching resistance of NHL2 and NHL5 mortars, but those of NHL3.5 have low confidence even though it is likely that the latter binder fits in the sequence between the other two.

All this has clear implications for the performance of NHL mortars subject to long term saturated conditions: higher hydraulicity binders perform better and carbonated mortars perform better than uncarbonated mortars. Based on the results presented here and in part 1 [10] the index of resistance to leaching, combined with the acceleration factor of 20 for the ammonium nitrate leachant established in [10], enables estimates to be made of the time to a certain depth of leaching by water and of the depth after a certain time of leaching. These estimates are of practical importance in field performance of lime mortars and some examples are given in Table 5. Table 5 shows the depth of leaching after exposure to water for one and 100 years, and the time taken to leach to a depth of 20 mm (taken as equivalent to a residual compressive strength of 25% – Table 4) and 50 mm (taken as half the depth of a typical mortar joint in the external face of a wall).

Table 5. Estimated leaching in NHL mortars exposed to water.

	NHL2 uncarbonated	NHL3.5 uncarbonated	NHL5 uncarbonated	NHL2 carbonated	NHL3.5 carbonated	NHL5 carbonated
Leached depth after one year (mm)	3.5	2	2.5	2	2	1.5
Leached depth after 100 years (mm)	33	22	24	20	20	15
Time to leach to 20 mm deep (years)	40	80	70	100	100	190
Time to leach to 50 mm deep (years)	230	500	440	630	630	1200

Reduction in mortar strength is not a serious structural problem in service because the compressive strength of a wall is dominated by the strength of the unit rather than the mortar. For example, structural design codes suggest that even when the mortar used with a given unit has a strength of only 1.5 MPa instead of 16 MPa the wall strength is reduced by only 50%, not the anticipated 90% [23]. However, using strength reduction as a proxy for loss of durability, dimensional stability, cohesion and adhesion of the mortar, all relationships that are well established for cementitious materials [24], it is clear that binder leaching will lead to erosion of the mortar bed and ultimately to loss of units from the masonry. The data in Table 4 confirm that any measures that will enhance the degree of carbonation of mortar will significantly prolong the service life of masonry.

The results from this work are generally consistent with the differences in solubility of $Ca(OH)_2$ and $CaCO_3$ but some points of detail will be explored next. The composition of the NHL2 binder [10] and the mortars (Table 1) suggest that the highest proportion of calcium in the hardened mortars tested is 42 g Ca/kg or 73.5 kg Ca/m^3 (the more hydraulic binders have lesser amounts of soluble Ca). In the accelerated leaching experiment 20 specimens (9 kg of mortar) were exposed to 15 L of leachant to produce $Ca(NO_3)_2$, whose solubility is 1440 g/L (351 g Ca/L). Even if all the Ca present in the mortar specimens were leached into solution by the end of the experiment this would deliver only 25.2 g Ca/L (7% of the saturation solubility level). This confirms that the experimental conditions have NH_4NO_3 in excess and that laboratory leaching is not solubility-limited. However, field leaching by water is a different matter. The solubility of $Ca(OH)_2$ is 1.6 g/L (0.86 g Ca/L) and of $CaCO_3$ is 0.0066 g/L (0.00264 g Ca/L). Thus 1 kg of mortar requires 26 L of water to dissolve the Ca if the latter were present as $Ca(OH)_2$and 6363 L of water if it were $CaCO_3$. Thus static leaching is clearly solubility-limited because it is improbable that mortar will be constantly surrounded by so much water. However, a different picture is presented by considering rainfall on unprotected masonry. This will pass through a 10 mm thick bed joint, estimated to contain 17.5 kg mortar per m^2. 17.5 kg of mortar contains 735 g Ca, which, if present as $Ca(OH)_2$, will dissolve in 854 L of water. Taking Scotland's rainfall as 1500 mm/year [25], corresponding to 1500 L/m^2surface area, this could take less than a year. Likewise, if the mortar were carbonated the $CaCO_3$ would require 185 years of rainfall. Because water from rain flows unevenly over surfaces and can be concentrated into particular areas, these figures may well be conservative. At the other extreme of climate related water flow, Hall et al. [4] calculated the upward water flow as a result of evaporative pumping in the London and Athens climates. For a 300 mm thick wall (with no damp course) with evaporation from both sides the annual water flow through the bed joint in a wall in London is 1380 L/m^2 and in Athens is 2633 L/m^2. These figures suggest that uncarbonated mortar is vulnerable to complete binder loss in less than one year and that even carbonated mortar in the Athens climate may be fully leached in 100 years. These calculations suggest that it is reasonable to use accelerated leaching experiments, where the leachant solution is in excess compared to the amount of Ca in the mortar, to indicate the scale of the binder leaching problem in field situations, where water is flowing at such a rate as to be also in excess. Stationary water is expected to be less of a problem.

Whereas the results presented here are a step towards a prediction model, further investigations are needed in order to establish the effects of variables such as water composition (e.g. Hardness and pH) and to study possible differences between 'site-leached' and 'accelerated-leached' mortars.

6. CONCLUSIONS

1. Molar ammonium nitrate solution has been shown to be a satisfactory leachant for accelerated deterioration studies on natural hydraulic lime binders in masonry mortar. It is able to extract calcium from laboratory specimens in a reasonable time for comparative studies of binder loss in carbonated mortar.

2. Leaching of calcium from the carbonated binder reduces the strength and increases the sorptivity of NHL mortar, with the kinetics following a first-order process. The rate constant increases from NHL5 to NHL3.5 to NHL2, i.e. as the lime binder becomes less hydraulic. This is consistent with carbonated NHL5 mortars having the highest leaching resistance.

3. The resistance to leaching of a given carbonated binder is higher than that of the same binder when uncarbonated. The experimental evidence suggests the following composite order of increasing resistance to leaching: uncarbonated NHL2 < uncarbonated NHL3.5 ≈ uncarbonated NHL5 ≈ carbonated NHL2 ≈ carbonated NHL3.5 < carbonated NHL5.

ACKNOWLEDGEMENTS

This investigation was financially supported by the Engineering and Physical Sciences Research Council under grant number EP/G064865/1 and by Historic Environment Scotland. We thank Rachel Hardie, Heath and Hardie Geosciences Ltd, for petrographic analysis, Bill Revie, Construction Material Consultants Ltd, for use of facilities for point counting, and Dr Clare Torney for helpful discussions. We additionally thank our project supporters: Roger Curtis, Historic Scotland, Paul Blackie, Jacobs, Steven Laing, Laing Traditional Masonry, Bryan Dickson, National Trust for Scotland, and Colin Tennant, lately of the Scottish Stone Liaison Group.

REFERENCES

1. M. Hulme, G.J. Jenkins, X. Lu, J.R. Turnpenny, T.D. Mitchell, R.G. Jones, Climate Change Scenarios for the United Kingdom: The UKCIP02 Scientific Report, Tyndall Centre for Climate Change Research, School of Environmental Sciences, University of East Anglia, 2002.

2. M. Cassar, C. Hawkins, Engineering Historic Futures: Adapting Historic Environments to Moisture Related Climate Change, Stakeholders Dissemination and Scientific Research Report, University College, London, 2007.

3. C. Hall, W.D. Hoff, Water Transport in Brick, Stone and Concrete, CRC Press, London, 2011.

4. C. Hall, A. Hamilton, W.D. Hoff, H.A. Viles, J.A. Eklund, Moisture dynamics in walls: response to micro-environment and climate change, Proc. R. Soc. A 467 (2011) 194–211.

5. S. Shrubsole, M. Macmillan, M. Davies, N. May, 100 unintended consequences of policies to improve the energy efficiency of the UK housing stock, Indoor Built Environ. 23 (2014) 340–352.

6. Sustainable Traditional Buildings Alliance, Responsible Retrofit of Traditional Buildings, STBA, London, 2012. (accessed May 2015).

7. A. McKinney, Would MOT Have Prevented Wall's Collapse?, The Herald, Glasgow, 2007 (August 29 2007)

8. A.M. Forster, Binder loss in traditional mass masonry: a cause for concern?, Struct Surv. 25 (2007) 148–170.

9. R. Kent, A heritage in ruins, The Building Conservation Directory, 2005, pp. 16– 19, (accessed February 2014).

10. A.M. Forster, E.M. Szadurski, P.F.G. Banfill, Deterioration of natural hydraulic lime mortars, I: effects of chemically accelerated leaching on physical and mechanical properties of uncarbonated materials, Constr. Build. Mater. 72 (2014) 199–207.

11. R.S. Boynton, Chemistry and Technology of Lime and Limestone, Wiley, New York, 1980.

12. V.H. Nguyen, H. Colina, J.M. Torrenti, C. Boulay, B. Nedjar, Chemo-mechanical coupling behaviour of leached concrete, Part 1: experimental results, Nucl. Eng. Des. 237 (2007) 2083–2089.

13. M. Berra, F. Carassiti, T. Mangialardi, A.E. Paolini, M. Sebastiani, Leaching behaviour of cement pastes containing nanosilica, Adv. Cem. Res. 25 (2013) 352–361.

14. BS EN 13139, Aggregates for Mortar, British Standards Institution, London, 2002.

15. BS EN 459-1, Building Lime: Definitions, Specifications and Conformity Criteria, British Standards Institution, London, 2010.

16. BS EN 459-2, Building Lime: Test Methods, British Standards Institution, London, 2010.

17. BS EN 1015-2, Methods of Test for Mortar for Masonry: Bulk Sampling of Mortars and Preparation of Test Mortars, British Standards Institution, London, 2002.

18. BS EN 196-1, Methods of Testing Cement: Determination of Strength, British Standards Institution, London, 2005.

19. M. Lawrence, Influence of binder and filler type on early strength development of lime mortars, Journal of Architectural Conservation 12 (1) (2006) 37–51.

20. S. Pavia, B. Toomey, Influence of the aggregate quality on the physical properties of natural feebly hydraulic lime mortars, Mater. Struct. 41 (2008) 559–569.

21. BS EN ISO 15148, Hygrothermal Performance of Building Materials and Products. Determination of Water Absorption Coefficient by Partial Immersion, BSI, London, 2001.

22. D.J. Graham, N.G. Midgley, Graphical representation of particle shape using triangular diagrams: an Excel spreadsheet method, Earth Surface Process. Landforms 25 (13) (2000) 1473–1477.

23. A.W. Hendry, S.R. Davies, B.P. Sinha, Design of masonry structures, Spon, London, 1997.

24. A.M. Neville, Properties of concrete, Prentice Hall, London, 2011.

25. Met Office, Temperature, Rainfall and Sunshine Time Series, Met Office, London, 2015. (accessed October 2015).

9

CHAPTER

Recycled Tyre Rubber Modified Bitumens for Road Asphalt Mixtures: A Literature Review☆

Davide Lo Presti

Nottingham Transportation Engineering Centre, University of Nottingham, Nottingham, UK

HIGHLIGHTS

- Introducing the Recycled Tyre Rubber (RTR) material as environmental problem as well as engineering resource.
- Introducing the wet process technology.
- Describing in details the existing products associated to the wet process high viscosity technology.
- Describing in details the wet process No Agitation technology.
- Comparing the described technologies and providing justifications and suggestions toward a widespread use of RTR-MBs.

ABSTRACT

Nowadays, only a small percentage of waste tyres are being land-filled. The Recycled Tyre Rubber is being used in new tyres, in tyre-derived fuel, in civil engineering applications and products, in moulded rubber products, in agricultural uses, recreational and sports applications and in rubber modified asphalt applications. The benefits of using rubber modified asphalts are being more widely experienced and recognized, and the incorporation of tyres into asphalt is likely to increase. The technology with much different evidence of success demonstrated by roads built in the last 40 years is the rubberised asphalt mixture obtained through the so-called "wet process" which involves the utilisation of the Recycled Tyre Rubber Modified Bitumens (RTR-MBs). Since 1960s, asphalt mixtures produced with RTR-MBs have been used in different parts of the world as solutions for different quality problems and, despite some downsides, in the majority of the cases they have demonstrated to enhance performance of road's pavement. This study reports the results of a literature review upon the existing technologies and specifications related to the production, handling and storage of RTR-MBs and on their current applications within road asphalt mixtures. Furthermore, considering that RTR-MBs technologies are still struggling to be fully adopted worldwide, mainly because of poor information, lack of training of personnel and stakeholders and rare support of local policies, the present work aims to be an up-to-date reference to clarify benefits and issues associated to this family of technologies and to finally provide suggestions for their wide-spread use.

Keywords: Recycled Tyre Rubber; Asphalt Rubber; Bitumen Rubber; Crumb Rubber; Terminal blend

1. TYRE RUBBER: ENVIRONMENTAL PROBLEM OR ENGINEERING RESOURCE?

The increasing number of vehicles on the roads of industrialised and developing nations generates millions of used tyres every year. About 1.4 billion tyres are sold worldwide each year and subsequently as many eventually fall into the category of end of life tyres (ELTs) (Figure 1). Moreover, the amount of ELTs in Europe, US and Japan are about to increase because of the projected growing number of vehicles and increasing traffic worldwide. These tyres are among the largest and most problematic sources of waste, due to the large volume produced and their durability. The US Environmental Protection Agency reports that 290 million scrap tyres were generated in 2003 (EPA, 2007). Of the 290 million, 45 million of these scrap tyres were used to make automotive and truck tyre re-treads. In Europe every year, 355 million tyres are produced in 90 plants, representing the 24% of world production [1]. In addition the EU has millions of used tyres that have been illegally dumped or stockpiled. The inadequate disposal of tyres may, in some cases, pose a potential threat to human health (fire risk, haven for rodents or other pests such as mosquitoes) and potentially increase environmental risks. Most countries, in Europe and worldwide, have relied on land filling to dispose of used tyres but the limited space and their potential for reuse has led to many countries imposing a ban on this practice. The current estimate for these historic stockpiles throughout the EU stands at 5.5 million tonnes (1.73 times the 2009 annual used tyres arising) and the estimated annual cost for the management of ELTs is estimated at € 600 million [2]. With landfills minimising their acceptance of whole tyres and the health and environmental risks of stockpiling tyres, many new markets have been created for scrap tyres.

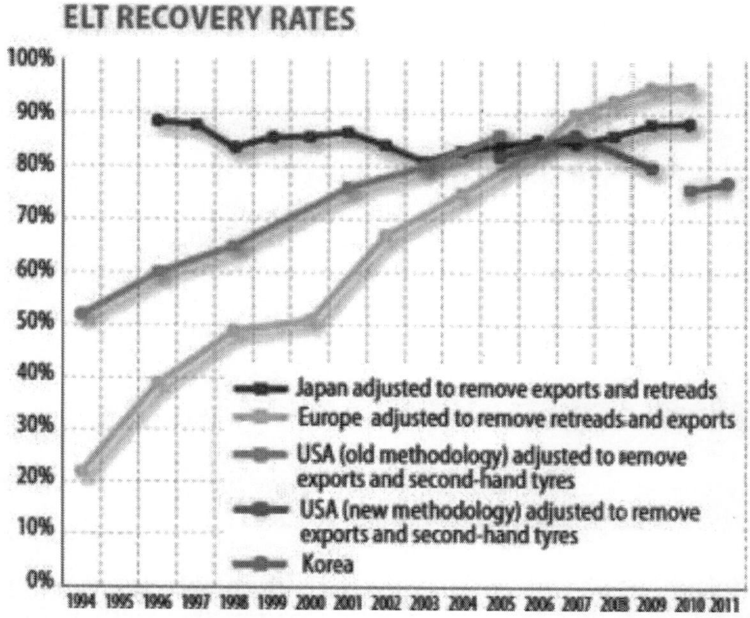

Figure 1. Evolution of ELTs recovery rates in major tyre markets, adapted from [2].

In order to face this problem, in Europe in 1989, a used tyres group composed of experts from the main tyre manufacturers producing in Europe, was set up under the strategic guidance of the European Tyre and Rubber Manufacturers Association (ETRMA). This Group was dedicated to the management of end of life tyres (ELTs). Also thanks to this group, since 1996, the collection rate has increased steadily while there has been a continuous decline in the land filling of used tyres (Table 1). In 2009 the European Union was faced with the challenge of managing, in an environmentally sound manner, more than 3.2 million tonnes of used tyres of which 95% were recovered. This confirms Europe as one of the most active areas in the world in the recovery of ELTs.

Table 1. ELTS management trends in the EU, adapted from [1].

	1994 (%)	1996 (%)	1998 (%)	2000 (%)	2002 (%)	2004 (%)	2006 (%)	2008 (%)	2008 (%)
Re-use/export	11	81	11	10	11	9	9	8	10
Reconstruction	10	12	11	11	11	12	11	10	8
Material recovery	6	11	18	19	25	28	34	38	40
Energy recovery	11	20	20	21	27	31	31	37	38
Landfill (disposal)	62	49	40	39	26	20	13	6	4

Country arise and recovery rates demonstrate that ELTs management in Europe is allowing the progressive elimination of land filling and raises the availability of Recycled Tyre Rubber (RTR) to be recycled for other purposes. In fact, the same characteristics that make waste tyres such a problem also make them one of the most re-used waste materials, as RTR is very resilient and can be reutilised in other products. These efforts should for example help to further develop the use of ELTs in rubberised asphalt in road construction, which has high growth potential in Europe and it is still relatively underutilised [2].

1.1. Recycled Tyre Rubber as engineering material

The tyre is a complex and high-tech safety product representing a century of manufacturing innovation, which is still on-going. From the material point of view the tyre is made up of three main components materials: (i) elastomeric compound, (ii) fabric and (iii) steel. The fabric and steel form the structural skeleton of the tyre with the rubber forming the "flesh" of the tyre in the tread, side wall, apexes, liner and shoulder wedge [3]. This engineering process is necessary to transform natural rubber in a product able to ensure performance, durability and safety. In fact, natural rubber is sticky in nature and can easily deform when heated up and it is brittle when cooled down (Table 2). In this state it cannot be used to make products with a good level of elasticity. The reason for inelastic deformation of not-vulcanised rubber can be found in the chemical nature as rubber is made of long polymer chains. These polymer chains can move independently relative to each other, and this will result in a change of shape. By the process of vulcanisation cross-links are formed between the polymer chains, so the chains cannot move independently anymore. As a result, when stress is applied the vulcanised rubber will deform, but upon release of the stress the rubber article will go back to its original shape. Compounding is finally used to improve the physical properties of rubber by incorporating the ingredients and ancillary substances necessary for vulcanisation, but also to adjust the hardness and modulus of the vulcanised product to meet the end requirement. Different substances can be added according to the different tyre mixtures; these include mineral oil and reinforcing fillers as carbon black and silica [4]. In general, truck TR contains larger percentages of natural rubber compared to that from car tyres [5]. Table 3 summarises the general tyre composition of tyres used in cars and trucks in the EU [2].

Table 2. Effect of temperature on natural rubber, adapted from [2].

−10 °C	Brittle and opaque
20 °C	Soft, resilient and translucent
50 °C	Plastic and sticky
120 °C −160 °C	Vulcanised when agents e.g., sulphur are added
180 °C	Break down as in the masticator
200 °C	Decomposes

Table 3. Comparison of passenger car and truck tyres in the EV, adapted from [3].

Material (contents)	Car (%)	Truck/buses (%)
Rubber/elastomers	48	43
Carbon black	22	21
Metal	15	27
Textile	5	–
Zinc oxide	1	2
Sulphur	1	1
Additives	8	6

From the structural point of view, the main components of a tyre are the tread, the body, side walls and the beads (Figure 2). The tread is the raised pattern in contact with the road. The body supports the tread and gives the tyre its specific shape. Beads are metal-wire bundles covered with rubber, which holds the tyre on the wheel. The inherent characteristics of the tyre are the same worldwide. They include: the resistance to mould, mildew, heat and humidity, retardation of bacterial development, resistance to sunlight, ultraviolet rays, some oils, many solvents, acids and other chemicals. Other physical characteristics include their non-biodegradability, non-toxicity, weight, shape and elasticity. However, many of the characteristics, which are beneficial during their on-road life as consumer products, are disadvantageous in their post-consumer life and can create problems for collection, storage and/or disposal [6].

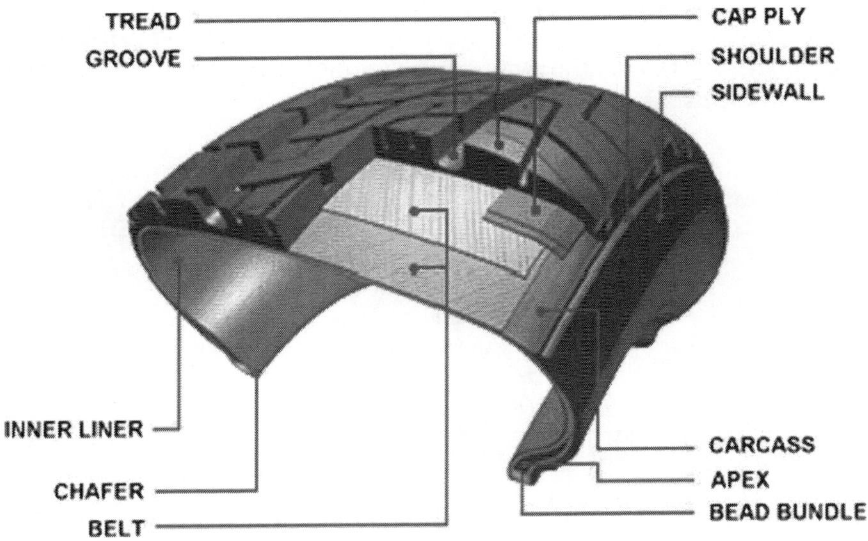

Figure 2. Tyre structure, adapted from [7].

1.2. From ELTs to Crumb Rubber Modifier

The tyre life cycle traditionally comprises five main stages, which includes extraction, production, consumption, collection of used tyres and waste management. A simplified version of the tyre cycle is illustrated in Figure 3. After the collection of ELTs, the next stage includes landfilling and recovery. Worldwide there has been a continuous decline in landfilling used tyres, while the recovery routes include different options such as: "energy recovery" where ELTs having a calorific value equivalent to that of good quality coal are used as an alternative to fossil fuels, or "chemical processing" such as pyrolysis, thermolysis and gasification, (the economic viability of these options has yet to be proved) and finally "granulate recovery". The latter involves tyre shredding and chipping processes which is carried out by using large machinery that cuts up tyres into small pieces of different sizes. At this stage, after the removal of the steel and fabric components, the material RTR can be used for a variety of civil engineering projects: rubberised asphalt pavements, flooring for playgrounds and sports stadiums, as shock absorbing mats, paving blocks, roofing materials, etc.

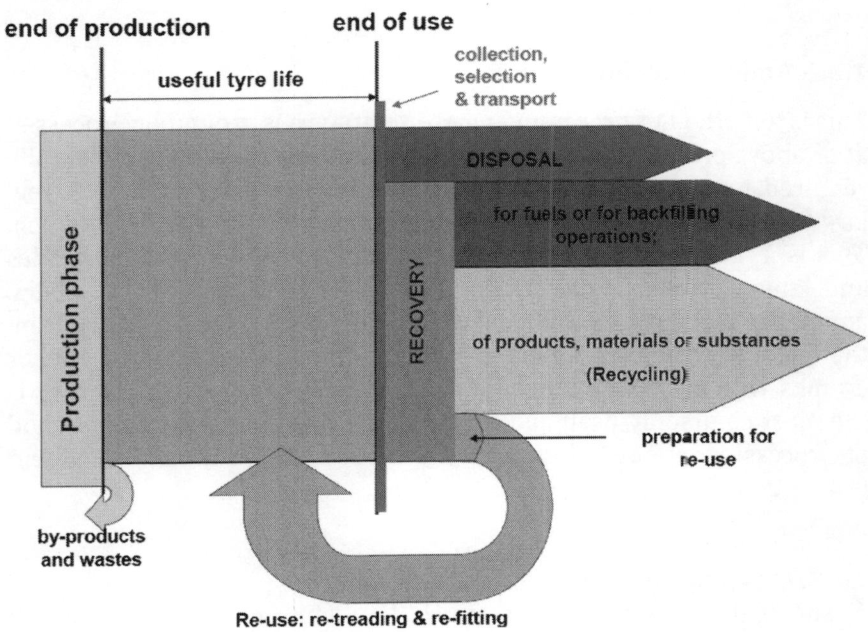

Figure 3. Life cycle of end of life tyres, adapted from [8].

The size of the tyre shreds may range from as large as 460 mm to as small as 25 mm, with most particles within the 100 mm to 200 mm range, while the tyre chips range from 76 mm down to approximately 13 mm. By further reducing the size of shreds and chips, it is possible to produce Ground and Crumb Rubber, also known as size-reduced rubber, which are suitable to be re-used in the asphalt industry. Crumb Rubber Modifier (CRM) is the common name used to identify the RTR particles used to modify bitumen. It is important to recognize that today CRM is typically a highly controlled material. The process is no longer just grinding up a stockpile of ELTs and adding the rubber to hot asphalt. The handling and shredding process is carefully planned and monitored to produce a clean and highly consistent rubber material. During the process, the tyre's reinforcing wire and fiber is removed. The steel is removed by magnets and the fiber is removed by aspiration. The resulting rubber particles are consistently sized and very clean. Automated bagging systems help ensure proper bag weights and eliminate cross contamination [9].

There are several technologies to reduce ELTs in CRM.

1.2.1. Ambient grinding

This is a method of processing where scrap tyre is ground or processed at or above ordinary room temperature. Ambient processing is typically required to provide irregularly shaped, torn particles with relatively large surface areas to promote interaction with the paving bitumen. This is a mechanical grinding, performed by means of rotating blades and knives, in which the critical step is the separation of the fibers, amongst which are generally included steel fibers. Once separated from the metallic material, ambient grinding is able to produce rubber crumbs with grain size ranging from 5 to 0.5 mm. Ambient grinding is the most commonly used and probably the most cost effective method of processing end of life tyres. A schematic representation of ambient grinding is shown in Figure 4.

1.2.2. Cryogenic grinding

As shown in Figure 5, this process uses liquid nitrogen to freeze the RTR (typically between −87 to −198 °C) until it becomes brittle, and then uses a hammer mill to shatter the frozen rubber into smooth particles with relatively lower surface area (Figure 6) than those obtained by ambient grinding (Figure 7 and Figure 5). Oliver [10] showed that several characteristics of the rubber granulate determine the elastic

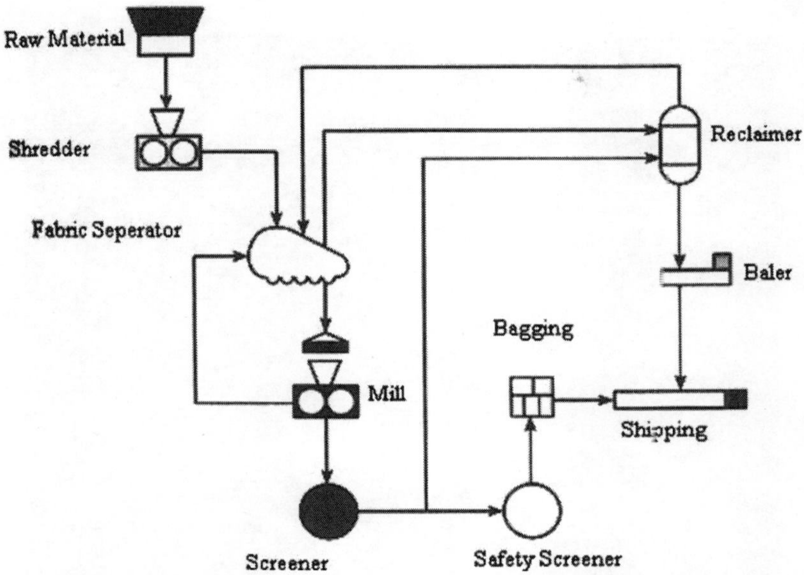

Figure 4. Schematic representation of ambient grinding, adapted from [3].

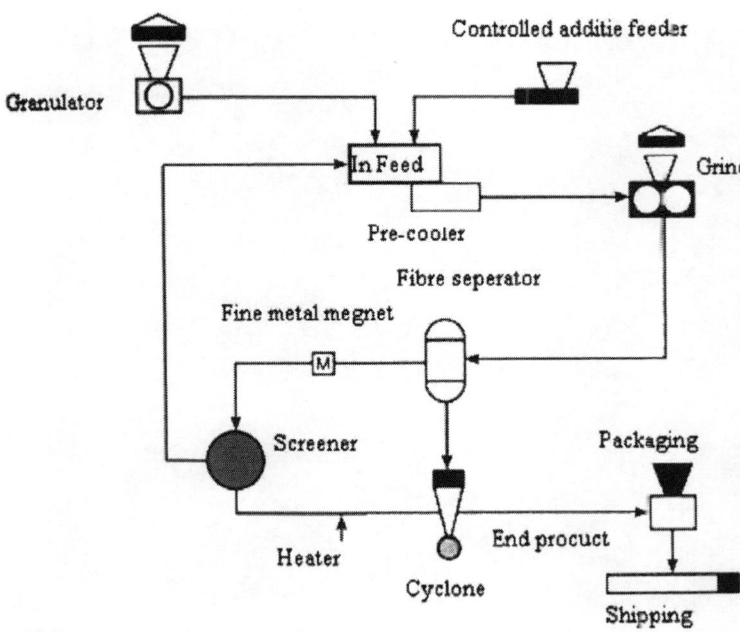

Figure 5. Schematic representation of cryogenic grinding, adapted from [3].

Figure 6. Ambient rubber crumbs. SEM analysis at 200um and 400x magnification, adapted from [12].

Figure 7. Cryogenic rubber crumbs. SEM analysis at 200um and 400x magnification, adapted from [12].

Table 4. Comparison between ambient and cryogenic ground rubber.

Physical property	Ambient	Cryogenic
Specific gravity	Same	Same
Particle shape	Irregular	Regular
Fibre content	0.5%	Nil
Steel content	0.1%	Nil
Cost	Comparable	Comparable

properties of the Crumb Rubber and those conferred on the final mix: they enhance with the decrease of specific gravity and particle size, and increase with the higher surface porosity of the granules. In fact, in wet process, rubber particles with a smooth surface, showed reduced reaction with the bitumen and worst the elastic properties of the mixture, if compared with those obtained by using granules with bigger porous surfaces and less specific weight. As a result, the use of CRM from cryogenic process in bituminous mixtures is discouraged [11]. A comparison between the properties of cryogenic and ambient ground rubber is summarised in Table 4.

1.2.3. Other processes

In addition to conventional ambient grinding techniques and the cryogenic process, there other less common proprietary processes currently in use to reduce RTR in crumbs or fine powder to be used as CRM:

- *Wet-grinding* is a patented grinding process in which tiny rubber particles are further size reduced by grinding into a liquid medium, usually water. Grinding is performed between two closely spaced grinding wheels. The obtained fine mesh RTR is mainly used as bitumen modifier [12].
- *Hydro jet size reduction*. This is a technique of processing RTR into finer particles with the help of pressurised water. Water jets at very high operating pressure (around 55.000 psi) rotate in high speed

arrays producing clean, wire-free rubbers crumbs (Figure 8). However, the process protocol is relatively new and still unknown to most in the industry. Nevertheless the high level of roughness of the resulting rubber crumbs makes this product very much attractive for bitumen modification. Figure 8 shows an example of a microscopic analysis of crumbs obtained through a patented hydro jet size reduction process.

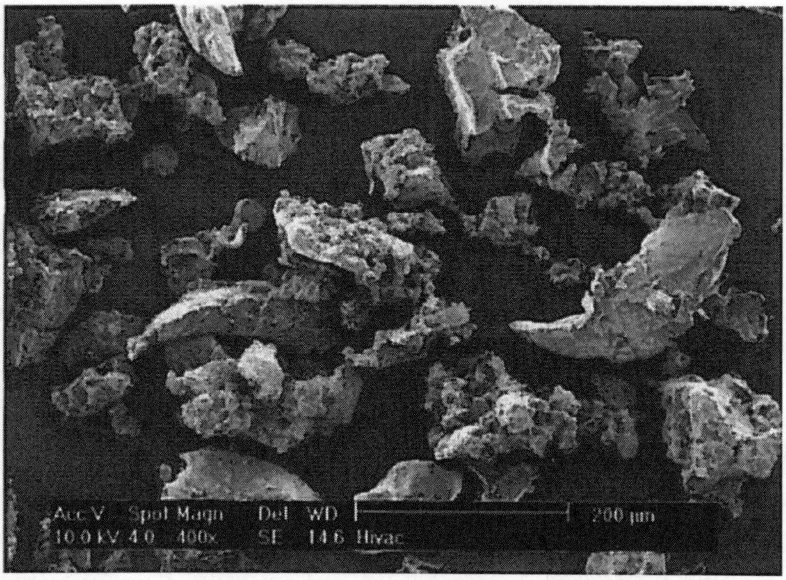

Figure 8. Hydrojet rubber crumbs. SEM analysis at 200um and 400x magnification, adapted from [12].

1.3. History of RTR in road asphalt mixtures

The accumulation of ELTs and premature pavement failures are both interconnected and dependant of each other due to enormous increase in traffic density and axle loading respectively. The use of RTR in asphalt pavements started 170 years ago, with an experiment involving natural rubber with bitumen in the 1840s [13], attempting to capture the flexible nature of rubber in a longer lasting paving surface. In 1960s scrap tyres were processed and used as a secondary material in the pavement industry. One application was introduced by two Swedish companies which produced a surface asphalt mixture with the addition of a small quantity of ground rubber from discarded tyres as a substitute for a part of the mineral aggregate in the mixture, in order to obtain asphalt mixture with improved resistance to studded tyres as well as to snow chains, via a process known as "dry process" [14]. In the same period Charles McDonalds, a materials engineer of the city of Phoenix in Arizona (USA), was the first to find that after thoroughly mixing crumbs of RTR with bitumen (CRM) and allowing it to react for a period of 45 min to an hour, this material captured beneficial engineering characteristics of both base ingredients. He called it Asphalt Rubber and the technology is well known as the "wet process" (Figure 9). By 1975, Crumb Rubber was successfully incorporated into asphalt mixtures and in 1988 a definition for rubberised bitumen was included in the American Society for Testing and Materials (ASTM) D8 and later specified in ASTM D6114-97. In 1992 the patent of the McDonald's process expired and the material is now considered a part of the public domain. Furthermore, in 1991, the United States federal law named "Intermodal Surface Transportation Efficiency Act" (then rescinded), mandated its widespread use, the Asphalt-Rubber technology concept started to make a "quiet come back" [15]. Since then, considerable research has been done worldwide to validate and improve technologies related to rubberised asphalt pavements.

Figure 9. Scheme of rubberised asphalt production through the "wet process".

Nowadays, these rubberised bitumen materials, obtained through the wet process, have spread worldwide as solutions for different quality problems (asphalt binders, pavements, stress absorbing lays and inlayers, roofing materials, etc.) with much different evidence of success demonstrated by roads built in the last 30 years.

2. RECYCLED TYRE RUBBER MODIFIED BITUMENS (RTR-MBS)

2.1. Geography of RTR-MBs in road asphalt mixtures

Since the invention of McDonald, the wet process technology has been used and modified more widely in four states in the US: Arizona, California, Texas, and Florida. More recently wet process has been used also in South Carolina, Nevada and New Mexico. The preference for using this particular modifier was due to the fact that not only does the utilisation of ELTs solve environmental problems but it also offers other benefits such as increased skid resistance, improved flexibility and crack resistance, and reduced traffic noise [16].

South Africa and Australia started introducing bitumen–rubber as a binder for asphalt and for seals from the early 1980s and mid 1970s respectively. In South Africa, both wet and dry processes were reported to have been used successfully although the dry process was mainly used in asphalt [17]. Two states in Australia (New South Wales and Victoria) adopted the wet process for limited application of rubberised asphalt, mainly as a crack resisting layer, but otherwise its usage has been predominantly for sprayed seal applications [18].

In Europe wet rubberised asphalt has been successfully used in road pavements application since 1981 in Belgium, as well as in France, Austria, Netherlands, Poland and Germany [19], more recently also in Greece [20] and UK [21], but the countries with a higher numbers of applications are Portugal [22], Spain [23], Italy [24], Czech Republic[25] and Sweden [26].

Nowadays the rubberised asphalt technology is being adopted in many other parts of the world: As reported by Widyatmoko and Elliot [18], Taiwan was reported to have adopted the Arizona DOT gap-graded and open-graded rubberised asphalt mixtures for flexible pavement rehabilitation; furthermore, rubberised asphalt has been trialled in Beijing and for use in new and maintenance work as part of the preparation for the 2008 Olympics in China and it has also been used in EcoPark Project in Hong Kong. On the basis of first positive experiences also Brazil [27] and Sudan [28] are strongly investing in the application of this technology for road pavements.

2.2. Overview of the bitumen – RTR interaction process

The term "wet process" refers to a whole family of technologies which varies a lot with regards with the chosen processing conditions. The nature of the mechanism by which the interaction between bitumen and CRM takes place has not been fully characterised. Traditionally it is reported that bitumen–rubber interaction is not chemical in nature [13], but other studies claim that the increase in binder viscosity cannot be accounted for only by existence of the rubber swelling particles [29].

The reaction itself is made up of two simultaneous processes (Figure 10): partial digestion of the rubber into the bitumen on one hand and, on the other, adsorption of the aromatic oils available in this latter within the polymeric chains that are the main components of the rubber, both natural and synthetic, contained in the RTR. The absorption of aromatic oils from the bitumen into the rubber's polymer chains causes the rubber to swell and soften [30].

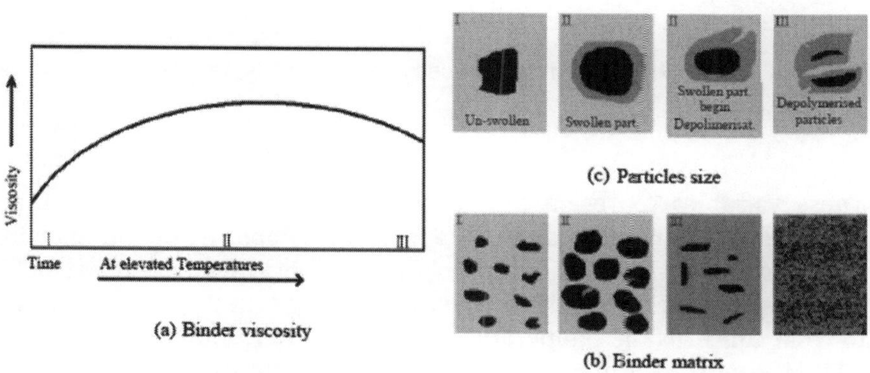

Figure 10. Bitumen-RTR interaction phenomenon at elevated temperatures: change of properties over time. Adapted from [31].

Rubber particles are swollen by the absorption of the bitumen oily phase at high temperatures (160–220 °C) into the polymer chains, which are the key components of the RTR-MB to form a gel-like material. Therefore, during the reaction there is a contemporaneous reduction in the oily fraction and an increase of rubber particle sizes with a consequent reduction of the inter-particle distance. This implies the formation of gel structures that produce a viscosity increase up to a factor of 10 [13].

Rubber reacts in a time–temperature dependent manner. If the temperature is too high or the time is too long, the swelling will continue to the point where, due to long exposure to the high temperatures, swelling is replaced by depolymerisation/devulcanisation which causes dispersion of the rubber into the bitumen. Depolymerisation starts releasing rubber components back to the liquid phase causing a decrease in the complex modulus (G^*), which is related to the material stiffness, while the the phase angle (δ), related to the elastic properties, continue to modify (Figure 10a and b). If temperature is high or time is long enough, depolymerisation will continue causing more destruction of the binder networking and so δ modification is lost [32]. The interaction between bitumen and rubber materials is material-specific and depends on a number of basic factors, including:

- Processing variables: temperature, time and device (applied shear stress).
- Base binder properties: bitumen source and eventual use of oil extenders.
- RTR properties: source, processing methods, particle size and content.

These variables represent the processing/interaction conditions that are necessary to monitor during the mixing of rubber within bitumen in order to govern the modification process. RTR-MBs are extremely dependent on the variability of these processing conditions, particularly to what concerns the temperature and time of reaction [33]. Moreover, RTR-MBs must be properly designed and, where necessary, produced to comply with specifications and provide a quality product suitable for the expected climate and traffic conditions. Depending on the adopted processing system, on the chosen processing conditions and on the selected materials, the wet process leads to different technologies as explained in the next sections.

2.3. Terminology associated with RTR-MBs

On the grounds of research done around the globe, rubberised bitumen is used as a general term to identify a group of concepts that incorporate RTR into bituminous binders for paving applications. These terms refer to the uses of RTR, in form of CRM as modifying agent in bitumen, that are different in their mix composition, method of production or preparation and in their physical and structural properties. The method of modifying bitumen with CRM produced from scrap tyres and other components as required before incorporating the binder into the bitumen paving materials, is referred to as the 'wet process'. Wet process is obtainable through two different processing systems.

2.3.1. McDonald process

This terminology is related to the system of producing RTR-MB with the original wet process proposed by Charles McDonald in the 1960s. The McDonald blend is a Bitumen Rubber blend produced in a blending tank by blending Crumb Rubber and bitumen. This modified binder is then passed to a holding tank, provided with augers to ensure circulation, to allow the reaction of the blend for a sufficient period (generally 45–60 min). The reacted binder is then used for mix production (Figure 11).

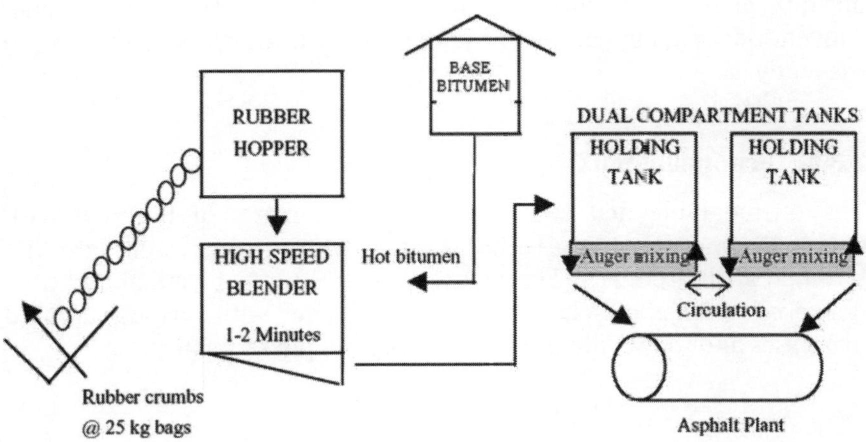

Figure 11. Schematic diagram of McDonald's wet process.

2.3.2. Continuous blending-reaction systems

This system is similar to the McDonald process of blending, the difference is that CRM and bitumen are continuously blended during the mix production or prepared by hand and then stored in storage tanks for later use. Therefore, it consists of a unique unit with agitators, in which the reaction occurs during the blending [13].

2.3.3. Field blends

Bitumen-RTR blends are typically produced at the asphalt plant by incorporating some modifications to the existing asphalt plant, for this reason they can be identified as field-blends. The above mentioned modifications include the introduction/adaptation of heated blending tanks, heated reaction tanks, rubber feed and storage tanks [13]. Hence, it is already few years that field-blended RTR-MBs can also be blended through less drastic modification to a standard hot mix asphalt plant through the portable production unit based on the McDonald system. The equipment is typically trailer mounted and is transported into the asphalt plant site. The blending unit receives ground rubber in the hopper, the rubber then moves to the mixing chamber to be blended with virgin liquid bitumen (Figure 12); the resulting rubberised bitumen is stored in the Portable Reaction Tank (Figure 13). Once reacted, binder is moved to a second compartment where it is fed to the hot-mix plant to be included in the HMA production. At the end, conventional paving equipment without modifications is used to place the material.

2.3.4. Terminal blends

The terminal blended rubberised binder consists of bitumen with Crumb Rubber Modifier (CRM) binder which is digested into the bitumen at the refinery (Figure 14) or at a bitumen terminal and then delivered to the plant. These binders are blended with various patented processes and they could present problems of phase separation.

Figure 12. Example of mobile Mixing unit for Asphalt-Rubber, adapted from [34].

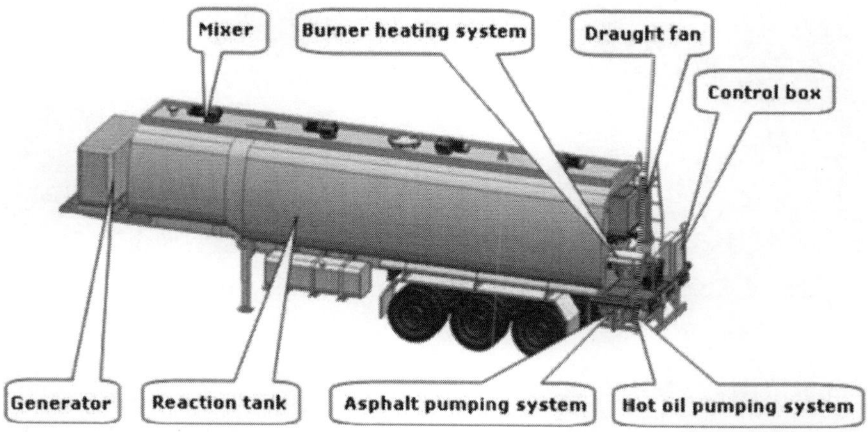

Figure 13. Example of mobile reaction tank for Asphalt-Rubber, adapted from [34].

Figure 14. Terminal blend binder system, adapted from [35].

2.3.5. Caltrans terminology

The most important distinction among the various rubberised bitumens seems to be related to the ability of obtaining a homogeneous RTR-MBs which do not present the issue of phase separation during storage periods. Based on this distinction, Caltrans [36]divides the wet process into two families linked to two very different types of RTR modification currently in use: the "wet process-high viscosity" and the "wet process-

No-Agitation". To promote a clear understanding, a detailed description of these technologies is provided in the next paragraphs.

3. Wet process-high viscosity

The original wet process-high viscosity, invented by Charles McDonald, leads to a product with a series of benefits which are basically all linked to the binder's increase in elasticity and viscosity at high temperatures that allows greater film thickness in paving mixes without excessive drain down or bleeding. According to Caltrans [36], the RTR-MB that maintain or exceed the minimum rotational viscosity threshold of 1500 cPs at 177 °C (or 190 °C) over the interaction period should be described as "wet process-high viscosity". These materials require continue agitation, with special equipment, to keep the RTR particles uniformly distributed. They may be manufactured in large stationary tanks or in mobile blending units that pump into agitated stationary or mobile storage tanks to be used directly at the job site. Phase separation is a big issue of these binders that needs therefore to be produced directly on the job-site or rarely transported to the field within tanks equipped with special augers, For this reason, these blends are often well-known as *Field blends.*

Wet process-high viscosity binders typically require at least 15% CRM to achieve the threshold viscosity. However, for some specifications [36] the viscosity requirements are meet also with less than 15% of RTR content. A number of RTR-MBs and mix specifications have been developed based on this original idea and each of them is related to a specific technology. This section will describe each of these technologies listed in Table 5 by providing a step by step framework to operate an appropriate plan production and storage of these materials. At last, this paragraph will provide an overview of the benefits and limitations related to the wet process-high viscosity.

Table 5. Existing specifications for rubberised asphalt mixtures.

Technologies	Specifications	Country
Asphalt-Rubber binder	ASTM D6114/D6114 M, [4] Caltrans' Asphalt-Rubber user guide, [5]	USA
Bitumen Rubber binder	SABITA Manual 19, [6] AsAc Technical Guideline, [7]	RSA
Crumb Rubber binder	Austroads, [8]	AUS and NZ

3.1. Asphalt-Rubber binder (USA)

According to the ASTM D6114 [37], Asphalt-Rubber is "a blend of asphalt cement, CRM and certain additives in which the rubber component is at least 15% by weight of the total blend and has reacted in the hot asphalt cement sufficiently to cause swelling of the rubber particles".

3.1.1. CRM requirements

According to ASTM D6114, in order to produce Asphalt-Rubber, the rubber should have the following characteristics:

- Less than 0.75% moisture and free flowing.
- Specific gravity of 1.15 ± 0.05.
- No visible nonferrous metal particles.
- No more than 0.01% ferrous metal particles by weight.
- Fibre content shall not exceed 0.5% by weight (for hot mix binder applications).
- Recommends all rubber particles pass the No. 8 (2.36 mm) sieve.
- (Note that Rubber gradation may affect the physical properties and performance of Bitumen Rubber hot mix).

Caltrans (2006) specifies that CRM to be used as modifier must include 25 ± 2% by mass of high natural rubber and 75 ± 2% of CRM from scrap tyre. Both types of rubber must meet specific chemical and physical requirements including gradation and limits on fabric and wire contaminants. The RTR consists primarily of 2 mm to 600 μn sized particles (No. 10 to No. 30 sieve sizes). The high natural rubber CRM is somewhat finer, mostly 1.18 mm to 300 urn sieve sizes (Table 6).

Table 6. Typical gradation of the RTR to be used in the Asphalt-Rubber binder, adapted from [38].

Sieve #	Nominal size (mm)	% Passing
10	2.36	100
16	1.18	75–100
30	600 μm	25–100
50	300 μm	0–45
100	150 μm	0–10
300	75 μm	0

3.1.2. Base binder requirements

ASTM D6114 specifies three different types of Asphalt-Rubber. Each of them is associated to a suitable bitumen to be used as base of modification

- Type-I binders typically include stiffer base bitumen and are generally recommended for hot climates, such as: AC-20, AR-8000 and PG64-16.
- Type-II binders typically include base bitumen softer than Type-I and are generally recommended for moderate climates, such as: AC-10, AR-4000 and PG58-22.
- Type-III binders typically include the softest grade base bitumen and are generally recommended for cold climates, such as: AC-5, AR-2000 and PG52-28.

where AC (asphalt cement) and AR (Aged Residue) are referred to the American grading systems based on viscosity. For example, an AC 20 asphalt has a viscosity of 2000 poise (+20%) at 60 °C, whilst an AR 4000 bitumen has a viscosity of 4000 poise (+25%) at 60 °C after aging. A rough comparison between penetration and viscosity (AC and AR) bitumen grades is shown in Figure 15.

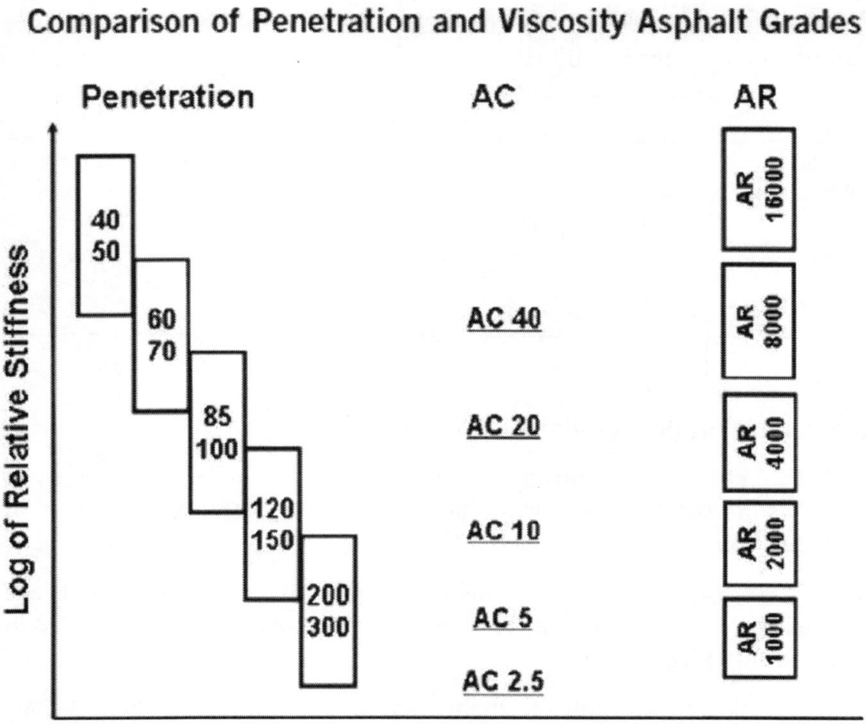

Figure 15. Comparison of Penetration and Viscosity bitumen grades.

3.1.3. Asphalt-Rubber plant production

By definition, Asphalt-Rubber is prepared using the "wet process-high viscosity" system. Physical property requirements are listed in ASTM D 6114, "Standard Specification for Asphalt-Rubber binder" (Table 8). The Asphalt-Rubber is produced at elevated temperatures (⩾350 F, 177 °C) in low shear (Shatanawi, 2010) to promote the physical interaction of the asphalt binder and rubber constituents, and to keep the rubber particles suspended in the blend. Various petroleum distillates or extender oil may be added, at a rate of 2.5–6% by mass of the bitumen binder, to reduce viscosity, facilitate spray applications, and promote workability. Field production of high viscosity Asphalt-Rubber binders is a relatively straightforward process and it is still based on the McDonald process (Figure 11). Equipment for feeding and blending may differ among Asphalt-Rubber types and manufacturers, but the processes are similar. The component materials are metered into high shear blending units to incorporate the correct proportions of extender oil and CRM into the paving grade asphalt. The blending units thoroughly mix the CRM into the hot asphalt cement and extender oil, and the blend is pumped into a heated tank where the Asphalt-Rubber interaction proceeds.

RTR is usually supplied in 1 ton (0.91 tonne) super sacks that are fed into a weigh hopper for proportioning. Augers are needed to agitate the Asphalt-Rubber inside the tanks to keep the CRM particles well dispersed, otherwise the particles tend either to settle to the bottom or float near the surface. Agitation can be verified by periodic observation through the top hatch or the port where the auger control is inserted. The Asphalt-Rubber binder must be interacted with agitation for a minimum of 45 min at temperatures from 190 to 218 °C to achieve the desired interaction between bitumen and rubber. In order to maintain the reaction temperature within the specified range, the bitumen must be hot, 204–224 °C before the design proportions of scrap tyre CRM and high natural rubber CRM are added. This is because the CRM is added at ambient temperature (not heated) and reduces the temperature of the blend. The component materials are metered into blending units to incorporate the correct proportions of CRM into the bitumen, and are thoroughly mixed. The Asphalt-Rubber producer is allowed to add the extender oil while adding the rubber, although in some cases the base binder may be supplied with the extender included If the Asphalt-Rubber producer adds the extender oil, use of a second meter is recommended to best control the proportioning. The meter for the extender oil should be linked to that for the bitumen. An Asphalt-Rubber binder interacted at lower temperatures will never achieve the same physical properties as the laboratory design, although it may achieve the minimum specification values for use. Hand held rotational viscometers are used to monitor the viscosity of the Asphalt-Rubber interaction over time for quality control and assurance. Before any Asphalt-Rubber binder can be used, compliance with the minimum viscosity requirement must be verified using an approved rotational viscometer (e.g. Brookfield). As long as the viscosity is in compliance and the interaction has proceeded for at least 45 min, the Asphalt-Rubber may be used [36].

Caltrans (2006) recognises that before starting the plant production, an appropriate Asphalt-Rubber binder design must be developed (Table 7). It also specifies that at least 2 weeks prior to start of construction the Contractor must supply to the Engineer, for approval, an Asphalt-Rubber binder formulation (design or "recipe") that includes results of specified physical property tests, along with samples of all of the component materials. Samples of the prepared Asphalt-Rubber binder must also be submitted to the Engineer at least 2 weeks before it is scheduled for use on the project.

Table 7. Caltrans specification for rubberised bitumen [36].

Test parameter	Specification limits
Apparent viscosity, Haake, 190 °C: cp	1500–4000
Cone penetration at 25 °C (ASTM D217): dmm	25–70
Softening point (ASTM D36): °C	52–74
Resilience at 25 °C (ASTM D3407): %	Minimum 18

Table 8. Specification for Asphalt-Rubber [37].

Binder specification	ASTM D6114 (2009)		
	Type I	Type II	Type III
Apparent viscosity 177.5 °C (ASTM D 2196): cp	1500–5000	1500–5000	1500–5000
Penetration at 25 °C (ASTM D5): dmm	25–75	25–75	50–100
Penetration at 4 °C (ASTM D5): dmm	Min 10	Min 15	Min 25
Softening point (ASTM D36): °C	Min 57.2	Min 54.4	Min 51.7
Resilience at 25 °C (ASTM D5329): %	Min 25	Min 20	Min 10
Flash point (ASTM D93): °C	Min 232.2	Min 232.2	Min 232.2
After TFOT (ASTM D1754), residual penetration at 4 °C (ASTM D5): %	Min 75	Min 75	Min 75
Climatic region	Hot	Moderate	Cold
Average minimum monthly temperature: °C	Min −1	Min −9	Min −9
Average maximum monthly temperature: °C	Min 43	Max 43	Max 27

3.1.4. Asphalt-Rubber storage

Caltrans requires heating to be discontinued if Asphalt-Rubber material is not used within 4 h after the 45-min reaction period. The rate of cooling in an insulated tank varies, but reheating is required if the temperature drops below 190 °C. A reheat cycle is defined as any time an Asphalt-Rubber binder cools below and is reheated to 190–218 °C. Caltrans allows two reheat cycles, but the Asphalt-Rubber binder must continue to meet all requirements, including the minimum viscosity.

Sometimes the binder must be held overnight. The bitumen and rubber will continue to interact at least as long as the Asphalt-Rubber remains liquid. The rubber breaks down (is digested) over time, which reduces viscosity. Up to 10% more CRM by binder mass can be added to restore the viscosity to specified levels. The resulting Asphalt-Rubber blend must be interacted at 190–218 °C for 45 min and must meet the minimum viscosity requirement before it can be used.

3.2. Bitumen Rubber binder (RSA)

Bitumen Rubber binder is a terminology used for the rubberised bitumen obtained by the wet process in South Africa. Bitumen Rubber binders combine rubber crumbs (Table 9) with bitumen at high temperatures to achieve a complex two phase product, named non-homogeneous binder. In the Technical Guideline of the South African Asphalt Acadamey[38], is also reported that properties of the modified binder used in hot mix asphalt will influence the engineering properties of the resultant mix, therefore the substitution of a conventional bitumen with a modified binder can result in higher air voids due to reduced workability of the higher viscosity modified binders. For this reason, it is important that split samples of the modified binder are sent to all the participating laboratories (e.g.: supplier, site and control laboratories) and tested before commencement of a project to ensure that the results are within the reproducible limits as prescribed by the test methods.

Table 9. Rubber crumbs requirements for bitumen–rubber, adapted from [38].

Property	Requirement	Test method
Sieve analysis (% mass)		MB-14
Passing screen (mm)		
1.180	100	
0.600	40–70	
0.075	0–5	
Poly-isoprene content (%m/m total hydrocarbon)	25 min	Thermo gravimetric Analysis
Fibre length (mm)	6 max	
Bulk density (g/cm^3)	1.10–1.25	MB-16

3.2.1. Rubber requirements

In South Africa, the Committee of Land Transport Officials (COLTO)'s specification requires the reclaimed rubber to have not less than 30% natural rubber by mass of hydrocarbon content [39], whilst the SABITA Manual [40] and the Technical Guideline of the South African Asphalt Acadamey [38] specify 60–75% natural rubber by mass of hydrocarbon content, with all rubber particles passing the 1.18 mm sieve. The CRM is produced by a mechanical size reduction process. CRM produced by cryogenic-mechanical techniques are not permitted in South Africa. The

CRM must be pulverised, free of fabric, steel cords and other contaminants.

3.2.2. Base bitumen requirements

The binder used in the production of the bitumen–rubber must be SABITA B12 or B8 road-grade bitumen (60/70 or 80/100 penetration-grade bitumen respectively) or a blend of these grades to provide a product with a particular viscosity and other prescribed properties.

3.2.3. Bitumen Rubber production

The binder is manufactured from blending penetration grade bitumen (72–82% by mass), plus extender oil (0–4%) plus rubber crumbs (18–24%) in a patented high shear mixer with a speed of 3000 r.p.m, at a temperature in excess of 180 °C but not more than 220 °C and for short periods before the introduction of rubber. During the addition of the rubber component, the blend cools down considerably and has to be re-heated to a temperature of 190–200 °C to ensure proper digestion of the rubber in the bitumen. Special manufacturing equipment is required to manufacture this highly viscous material. The product has a limited useable life of 4–6 h and, therefore, manufacture usually takes place onsite, or very close to the construction site. Bitumen Rubber binder can be used for surface dressing operations, in which case it is applied with binder distributors specially designed to handle this highly viscous binder. For surface dressing applications, the bitumen–rubber binder is manufactured using the "wet process", which is also the most used for the manufacture of bitumen–rubber hot-mix asphalt. The blending unit consists of a small tank equipped with a high speed stirring device that ensures proper "wetting" of the rubber by the binder and prevents the formation of rubber lumps in the final product. From the blending tank the product is transferred to a digestion tank which could also be a specialised binder distributor. In the digestion tank the product is continually agitated while being heated to the final temperature.

The ratio of components varies depending on the bitumen source, the climatic conditions and the application. The more reliable manufacturers in South Africa nowadays prefer to standardise on 20% rubber content and also to preselect the type of tyres for the modification process. Following the addition of the rubber, a digestion period is required for the rubber to swell and partially dissolve in the bitumen/extender oil blend. The rubber never completely dissolves in the bitumen and the product is thus classed as a non-homogenous binder. The Bitumen Rubber blend is then circulated in a holding tank and heated at high temperatures (190–210 °C) to facilitate the chemical digestion process.

The extender oil could either be added to the penetration grade bitumen before delivery or to the bitumen on site. The requirements of the extender oil are such that it should have a flash point of greater than 180 °C and the percentage by mass of aromatic unsaturated hydrocarbons be greater than 55. To prevent sticking of rubber particles, also an addition of calcium carbonate or talc up to 4% by mass of rubber is permitted. On completion of the digestion period, a hand-held viscometer is used to perform a viscosity test on the product to confirm that sufficient digestion has taken place (Table 10). If approved, the product is ready for application [38].

Table 10. Properties of Bitumen Rubber surfacing seals and asphalt, adapted swfrom [38].

Property		Unit	Test method	Class	
			S-R1	A-R1	
Softening point[1]	°C	MB-17	55–62	55–65	
Dynamic viscosity @ 190 °C	dPa s	MB-13	20–40	20–50	
Compression and recovery	5 min		MB-11	>70	>80
	1 h	%		>70	>80
	4 days			>25	n/a
Resilience @ 25		%	MB-10	13–35	13–40
Flow		Min	MB-12	15–70	10–50

3.2.4. Bitumen Rubber storage

Bitumen Rubber degrades rapidly at application temperatures which are in excess of 200 °C. Therefore the blending of Bitumen Rubber generally takes place in close proximity to the spray site or asphalt mixing plant. On completion of the digestion period, the product generally has a further useable life at the application temperature of approximately 4 h. The rate of degradation will vary depending mainly on the application temperature and can be monitored on-site with a hand held viscometer. The manufacturer of the Bitumen Rubber should supply temperature curves showing the changes in the properties over time.

Figure 16 shows typical changes in the viscosity properties of a Bitumen Rubber at different temperatures over time. Only sufficient quantities of Bitumen Rubber should be blended at any time in accordance with what can be sprayed or mixed within the application viscosity window of the product. Allowance must be made for changing weather conditions and construction delays. Proper planning and close cooperation between the supplier and contractor is essential to limit the over production of Bitumen Rubber which may result in unnecessary degradation of the Bitumen Rubber over prolonged periods of heating. Product must not be superheated if it is not going to be used. This will enable the product to be superheated at a later stage for reuse if it is still within specification. If it happens to be out of specification then only 25% of the product can be re-blended with new bitumen and RTR. Table 11 shows the recommended temperatures and time limits for the short-term handling, storing, spraying, mixing and application binders modified with Bitumen Rubber. S-R1 is a Surface Seal, A-R1 is for hot mix while C-R1 is a Crack sealant and all of them are Bitumen Rubber binders.

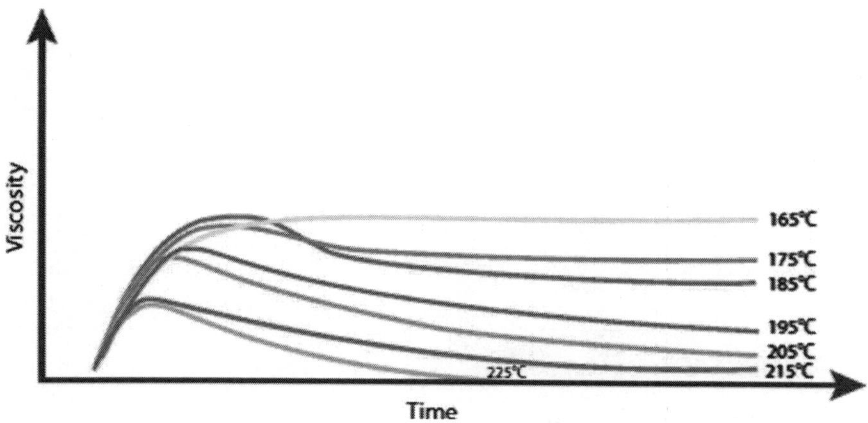

Figure 16. Typical changes in viscosity values from Bitumen Rubber at different temperatures over time, adapted from [38].

Table 11. Typical temperature/time limits for Bitumen Rubber, adapted from [38].

Binder class	Short term handling		Storage		Spraying/asphalt mixing/application		
	Max temp (°C)	Max holding time (h)	Max temp (°C)	Max holding time (h)	Max temp (°C)	Min temp (°C)	Max holding time (h)
S-R1	165	24	150	240	210	195	Refer to time/viscosity curve
A-R1	165	24	150	240	210	190	
C-R1	165	24	–	–	190	180	–

3.3. Crumb Rubber Modified binder (AUS & NZ)

Bitumen technologists in Australia have been looking into using rubber in bitumen since the early 1950s, when a surface dressing trial utilising 5% vulcanised reclaimed rubber in R90 bitumen (100 pen) with 10 mm aggregate was applied on the Prince Highway at St. Peters in New South Wales. However, there was not much development until the mid-1970s, when the Road authority of the state of Victoria (VicRoads) in conjunction with the Australian Road Research Board (ARRB) started using Crumb Rubber Modified bitumen in spray seal in urban and rural applications. It has been used extensively as a bitumen modifier in sprayed bituminous seals and occasionally in bitumen applications in Victoria. Thanks to this experience VicRoads, in collaboration with the Roads and Traffic Authority (RTA), New South Wales and the Main roads by Western Australia, have prepared the "Scrap rubber bitumen guide".

Since 1986, success has been reported using rubberised bitumen in Australia, mainly RTA and VicRoads, specifically when used as a crack resisting layer, e.g. thin bitumen overlay over concrete, Stress Absorbing Membranes or Stress Absorbing Membranes Interlayers. In 2006, Austroads published technical specifications framework AP-T41 coupled with a guide to the selection and use of Polymer Modified Binders: AP-T42 including the use of rubberised bitumen, identified as Crumb Rubber Modifed binder [41]. The reports suggest a wide usage of rubber into asphalt with both wet and dry process. The AP-T42 guide includes also a "Crumb Rubber Protocol" in which Constituent Specification, Binder Design, Field Blending, Binder Specification Limits and other restrictions regarding and Sampling and Testing to check requirements are indicated. An important component of this Crumb Rubber specification is the role of laboratory testing as explained later.

3.3.1. CRM requirements

In Austroads (2006), two levels of rubber content, nominally 15 and 18 are specified for Crumb Rubber Modified binder obtained through the wet process. Another level, from 25 to 30% , is designed for the dry process. Crumbs have only 2 sizes, namely 16 (coarse) and 30 (fine). Bulk density (max 350 kg/mc), grading, maximum particle size, steel content (<0.1%), moisture (<1%) and other properties must be specified by the contractor. All CRM shall be less than 3 mm in length and the rubber shall not contain any foreign material such as sand, fibre or aggregate. All this info is summarised in Table 12.

Table 12. Austroads specification for Crumb Rubber for bitumen modification, adapted from [41].

Test	Methods	Size 16	Size 30
Grading			
Passing 2.36 mm	AG/PT 1:43	100	100
Passing 1.18 mm	AG/PT 1:43	80 (min)	100
Passing 0.60 mm	AG/PT 1:43	10 (max)	40 (max)
Passing 0.15 mm	AG/PT 1:43		5 (max)
Bulk density	AG/PT 1:44	350 kg/m^3	350 kg/m^3
Water content	AG/PT 1:43	<1%	<1%
Steel content	AG/PT 1:43	<0.1%	<0.1%

3.3.2. Base bitumen requirements

The only requirements related to the most suitable bitumen to be used as base for the Crumb Rubber Modified binders, is indicated in the "Crumb Rubber Protocol" of the technical guide of Austroads [41], which specifies to use Bitumen Class 170 corresponding to a Penetration grade of 85/100 (Table 13).

Table 13. Summary of the most important specification requirements for RTR-MBs.

Properties	ASTM 2009			Caltrans 2006	AsAc 2007	Austroads 2007
	Type I	Type II	Type III			
Base bitumen requirements						
Penetration: dmm	85–100	120–150	200–300	120–150	60–100	85–100
Crumb Rubber requirements						
Passing sieve: mm	2.36			2.36	1.18	2.36
Rubber content: %	⩾15			18–22	18–24	15–18
Additives						
Extender oils: %	–			2.5–6	0–4	–
Caclium carbonate /Talc: %	0–4			–	0–4	–
Processing conditions						
Temperature: °C	177			190–220	180–220	195
Mixing speed: rpm	–			–	3000 rpm	–
Mixing time: min	45 + reaction			45–60	–	30–45

3.3.3. Crumb Rubber Modified binder production

As mentioned previously, the Austroads guide for Selection and use of Polymer Modified Binders AP-T42 [41], gives a major importance to the a priori laboratory testing. In fact, the design process requires a laboratory exercise to be undertaken using the following procedure:

1. A trial mix is prepared for each of an appropriate series of rubber concentrations such that the specification limits for the selected binder grade are met in at least one of the mixes (mixing temperature 195 °C, digestion period 45 min for size 16 rubber and 30 min for size 30).

2. The measured properties are plotted against rubber concentration and the concentration at which the specification limits are complied with is identified. This rubber concentration is deemed to be the design concentration.

3. A rubber extraction is performed on the design mix using AG:PT/T1 42 Laboratory method for determining Crumb Rubber concentration. The determined concentration of rubber is used as a check for the analysis of field collected samples.

The diagram in Figure 17 illustrates the key steps. In the Austroads 2006 specifications, S40R, S45R/S50R and S55R were replaced by S15RF, S18RF, for asphalt application, and S55R was kept for sprayed sealing. Plant protocols may require different procedures to the field defined system and the contractor will be responsible for the binder design. Crumb Rubber Modified binders are produced with production routes: factory and field produced rubberised bitumen.

- *Factory produced Crumb Rubber Modified binders*: usually include combining oil and lower concentrations of rubber than their field produced counterpart. Factory produced Crumb Rubber Modified binder are used for hot mix asphalt by using the wet process. This has been shown to improve the mix properties but requires relatively high binder contents. Since the concentration of rubber is an important performance controlling factor, generally the higher the volumetric concentration, the more effective will be the binder in more demanding applications. Flux oil should only be added to the CRM classes and the manufacturer's recommendations should be followed. It is important to use the finer CRM in order to achieve a reasonable level of digestion in the binder in the short time that the material remains hot.

- *Field produced Crumb Rubber Modified binders*: are produced for sprayed sealing. They are a high temperature blend of bitumen and CRM without combining oils, and may contain up to 25% by mass of rubber; these formulations are not suitable for long distance transportation or extended storage but represent the highest performing materials within their class when optimally digested. The properties of the field produced rubberised bitumen are generally controlled by their softening point, elastic recovery and/or torsional recovery [41].

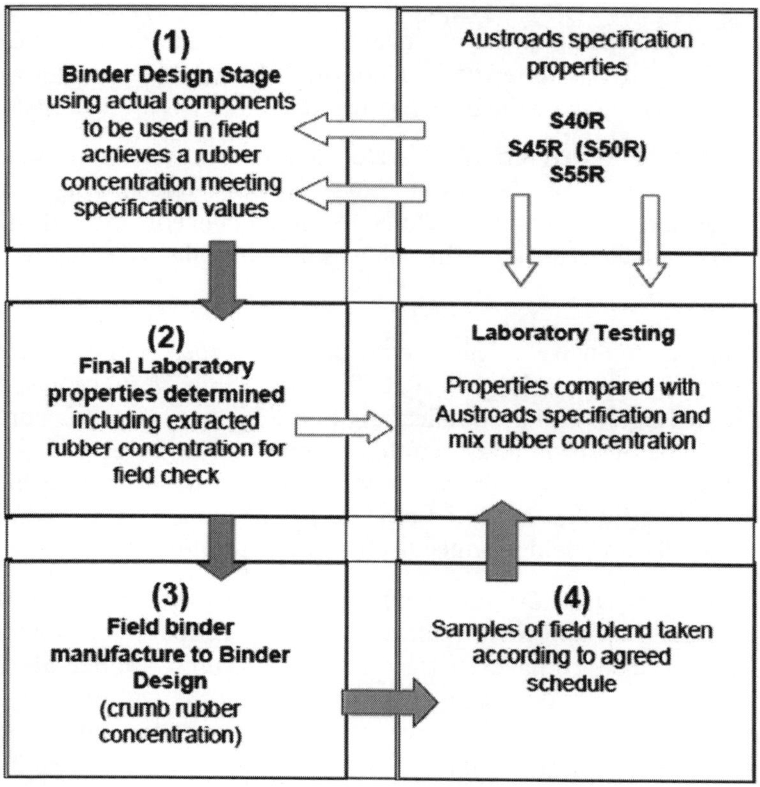

Figure 17. Austroads laboratory procedure for Crumb Rubber modified binders, adapted from [41].

3.3.4. Crumb Rubber Modified binder storage

When field mixing Crumb Rubber binders, the material must be continually circulated to minimise settling out of any rubber particles. Failure to do this will most likely result in blockages of the spraying jets and/or pipework. Do not store field produced Crumb Rubber mixtures in bitumen sprayers, road tankers or bulk storage because of the potential problem of segregation and settling out of the rubber particles resulting in blocked pipework etc.

It is worth to highlight that current work in Australia is focussing on the use of Crumb Rubber as a substitute for other modified binders in more conventional applications while using similar design criteria, with particular interest in the use of pre-blended products which behave much more like other pre-blended modified binders. Laboratory investigations with pre-blended Crumb Rubber Modified binders indicate satisfactory performance but at higher binder contents (>8%).

Table 13 summarises some of the most important specifications requirements for the RTR-MB production around the world.

3.4. Wet process-high viscosity: benefits, issues and limitation

Extensive literature clearly shows the numerous successes obtained using rubberised bituminous mixtures produced with the wet process-high viscosity. However, this technology is not a panacea. This section was thought to explain the main benefits provided by the usage of this technology and the issues that still exist and, in some cases, limit its extensive application.

3.4.1. Benefits

The primary reason for using RTR-MBs is that it provides significantly improved engineering properties over conventional paving grade bitumen. As for Asphalt-Rubber binders, they can be engineered to perform in any type of climate as indicated in ASTM D 6114. At intermediate and high temperatures, the RTR-MB shows physical and rheological properties significantly different than those of neat paving grade bitumens. The rubber stiffens the binder and increases elasticity (proportion of deformation that is recoverable) over these pavement operating temperature ranges, which decreases pavement temperature susceptibility and improves resistance to permanent deformation (rutting) and fatigue with little effect on cold temperature properties [36].

As demonstrated by various researchers, RTR-MBs obtained through the wet process have reduced fatigue and reflection cracking, greater resistance to rutting, improved aging and oxidation resistance and better chip retention due to thicker binder films [42],[43], [44], [45], [39] and [46]. Also Asphalt-Rubber pavements have been demonstrated to have lower maintenance costs [47], lower noise generation [48], [49] and [50] higher skid resistance and better night-time visibility due to contrast in the pavement and stripping[51]. A summary of the benefits of using High-viscosity RTR-MBs for road pavement applications is given below:

High Viscosity RTR-MBs have:

- Increased viscosity that allows greater film thickness in paving mixes without excessive drain down or bleeding.
- Increased elasticity and resilience at high temperatures.

High Viscosity RTR-MBs pavements have:

- Improved durability.
- Improved resistance to surface initiated and fatigue/reflection cracking due to higher binder contents and elasticity.
- Reduced temperature susceptibility.
- Improved aging and oxidation resistance due to higher binder contents, thicker binder films, and anti-oxidants in the RTR.
- Improved resistance to rutting (permanent deformation) due to higher viscosity, softening points and resilience (stiffer, more elastic binder at high temperatures).
- Lower pavement maintenance costs due to improved pavement durability and performance.

In addition, High Viscosity RTR-MBs pavements and binders can result in:

- Reduced construction times due to thinner lifts.
- Reduced traffic noise (primarily tyre noise).
- Improved safety due to better long-term colour contrast for pavement markings because carbon black in the rubber acts as a pigment that keeps the pavement blacker longer.
- Savings in energy and natural resources by using waste products and not contributing to the stockpiles.

3.4.2. Limitations

Wet process-high viscosity materials are useful, but they are not the solution to all pavement problems. The Asphalt-Rubber materials must be properly selected, designed, produced, and constructed to provide the desired improvements to pavement performance. Pavement structure and drainage must also be adequate. Limitations on the use of Asphalt-Rubber include [36]:

- High Viscosity RTR-MBs are not best suited for use in dense-graded HMA. There is not enough void space in the dense-graded aggregate matrix to accommodate sufficient rubberised binder content to enhance performance of dense-graded mixes enough to justify the added cost of the binder.
- Construction may be more challenging, as temperature requirements are more critical. Asphalt mixtures with High Viscosity RTR-MB must be compacted at higher temperatures than dense-graded HMA because, like polymers, rubber stiffens the binders at high temperatures. Also, coarse gap-graded mixtures may be more resistant to compaction due to the stone nature of the aggregate structure.
- Potential odour, also if it seems to not be harmful (see environmental issues).
- It is not possible to store High Viscosity RTR-MB at elevated temperatures without equipping storage tanks with augers or paddles. Furthermore,
- these binders cannot be stored for prolonged period. If work is delayed more than 48 h after blending the High Viscosity RTR-MB may not be usable. Takallou and Sainton [52], said "the rubberised binder must be used within hours of its production". The reason is that if over-processed the CRM will be digested to such an extent that it is not possible to achieve the minimum specified viscosity even if more CRM is added in accordance with specified limits [37].

Economic Issues related to the usage of RTR-MBs have been adequately addressed by a large number of research projects and reports and also through long standing construction evaluation. For the sake of brevity, the most notable are discussed below:

Higher initial costs. Costs are higher than conventional bitumen per unit ton until economies of scale are in place. From a study conducted in the USA by Hicks and Epps[53], it has been experienced that the Asphalt-Rubber hot mix could costs almost double than conventional mixes. Nevertheless, since the year 2000 a falling cost difference trend has been registered, also because costs of construction materials and petroleum products have increased (Figure 18). Given this change in the cost structure it is easy to observe that High Viscosity RTR-MB is presently very attractive in cost when particularly examined in light of actual usage. With the RTR being a very cheap waste material the RTR-MBs technologies are increasing their appeal also from the economical point of view.

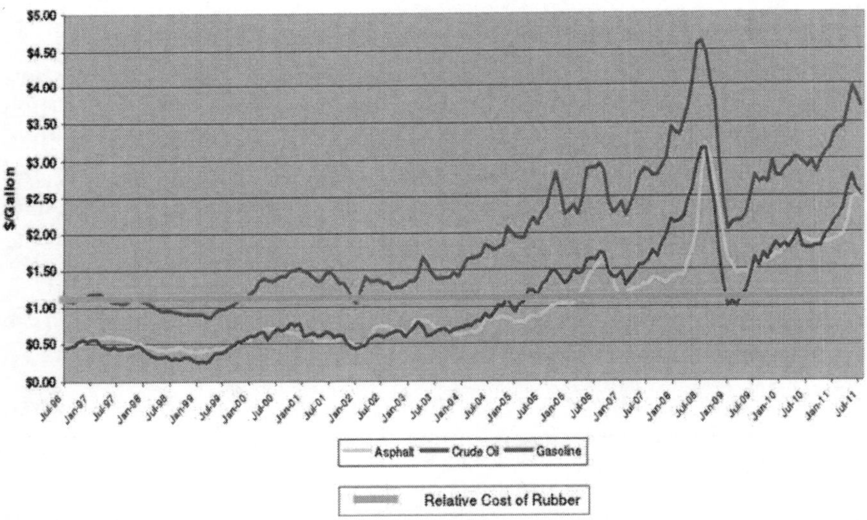

Figure 18. Crude oil, Fuel, Bitumen and rubber cost's trends, adapted from [54].

Lifecycle economics. As in any economical evaluation, cost effectiveness should be evaluated using Life Cycle Cost Analysis. Again, Hicks and Epps (2003) showed that evaluating different scenarios, in terms of pavement design, maintenance and rehabilitation strategies, the following was concluded:

- High Viscosity RTR-MB (Asphalt-Rubber) is a cost effective alternate for many highway pavement applications mainly due to the ability to reduce thickness when using rubberised asphalt. But also that,

- High Viscosity RTR-MB (Asphalt-Rubber) was not cost effective in all applications.
- When variability is considered in the inputs (cost, expected life, etc.), the Asphalt-Rubber alternates would be the best choice in most of the applications considered.

Other studies conducted at the Arizona State University [47] compared maintenance and user costs trends for the conventional bituminous pavements and Asphalt-Rubber pavements (Figure 19). Results showed that after 5 years the maintenance and user costs are not much different, after 10 years the maintenance cost begins to substantially be different, as higher maintenance costs will be anticipated for the conventional pavement. This difference for user costs starts at about 15 years. Based on the data analysis presented for the two pavements, an Asphalt-Rubber pavement would be more cost-effective than a conventional pavement with respect to road authorities costs as well as user costs.

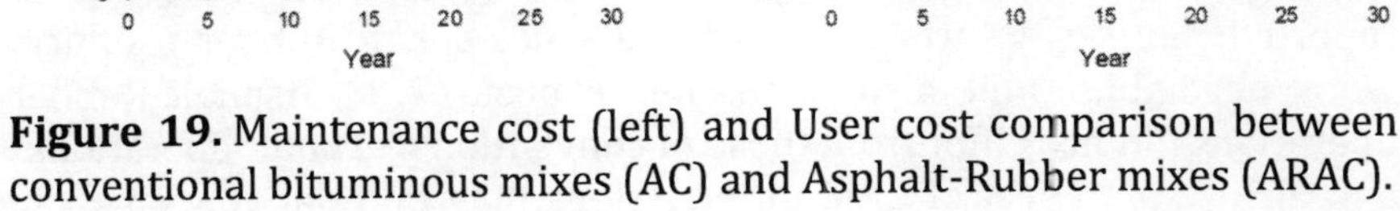

Figure 19. Maintenance cost (left) and User cost comparison between conventional bituminous mixes (AC) and Asphalt-Rubber mixes (ARAC).

Plant modifications. Another issue that contribute to higher initial cost of the wet process-high viscosity are plant modifications. In case of the production of field blends it is necessary to adapt standard hot mix asphalt plant, for instance with portable units as illustrated in Section 2.3.3. Additionally, conventional paving equipment without modifications is used to place the material. From an investigation made in 2007 in UK by the Waste & Resources Action Programme [18] in order to study a possible adoption of this technology, resulted that the approximate cost for importing/hiring and running the blending plant for 5 to 7 days of rubberised bitumen production is estimated to be between £60,000 and £90,000 (70.000–105.000€). This is because none special mixing units described above is currently available in the UK, and therefore requires importing/hiring from either the US or mainland Europe. Moreover, for the installation of a Portable rubberised bitumen mixing plant, an area of not less than 150 m^2 will be required, depending upon the model and make of the mobile plant [18]. It is worth highlighting that mobilisation and set up of the field-blends binder production equipment cost as much for small jobs as for big ones, but large projects spread mobilisation costs over more asphalt tonnage. The memo suggests that smaller projects may not be cost effective with respect to initial cost. Although the break point for project size may have changed since then, unit costs of small projects (three days paving or less) should be evaluated by LCCA during the design phase [36].

3.4.3. Environmental Issues

High Viscosity RTR-MBs provides the obvious environmental benefit of using waste tyres. Nevertheless, their production and their applications in asphalt presents the following issues:

Hazardous Emissions. Fume emissions have been studied extensively in a number of Asphalt-Rubber projects in USA since 1989. Different studies, performed also by the National Institute for Occupational Safety and Health (NIOSH), Federal Highways Administration (FHWA), [55], determined that use of Asphalt-Rubber does not appear to increase health risks to paving personnel. In these studies the following is reported: "...risks associated with the use of Asphalt-Rubber products were negligible" and also "Emission exposures in Asphalt-Rubber operations did not differ from those of conventional asphalt operations" and also "...the effect of CRM on emissions may be relatively small in comparison to the effects of other variables" [36]. Those variables include the fueling rate of the dryer, mix temperature, asphalt throughput rate and asphalt binder content.

However, some agencies also concluded that the rubber modified mix had an objectionable odour [21]. Moreover, in Colorado while researching and developing specifications for the use of rubberised asphalt for the 2006 paving season, local contractors expressed concerns about using the "wet or dry methods", due to the excessive smoke and smell that would be expelled into the atmosphere during the manufacturing process of this material at their asphalt plants. The contractors were so concerned about losing their state environmental certifications that they indicated they would not use the wet or dry methods without some assurances that their operating permits would not be jeopardised [56]. Nevertheless, several recent studies coupling the RTR-MBs with the Warm mix technology have shown a significant reduction of emission during field operations due to an important decrease of the mixing and compaction temperatures [57] and [21]. Therefore, environmental concerns upon the widespread usage of the wet process still exist, but the development of these technologies is proving that also this issue can be significantly reduced.

Recyclability. Before 1992, in USA Asphalt-Rubber pavements had been performing well and the replacement/recycling of them was not necessary. As some sections of Asphalt-Rubber pavements have met their service life span, recyclability became an issue. Carlson and Zhu [58] report of two jobs occurred in the USA where Asphalt-Rubber mixes were successfully recycled. One example is the recycling job occurred in the City of Los Angeles, California [59], where the initial placement of the Asphalt-Rubber pavement occurred in 1982. In 1994 the pavement was milled and stockpiled at a nearby asphalt plant. The Asphalt-Rubber grindings were added to the virgin rock and oil so that the grindings composed 15% of the final mix. At another location, the grindings were put through a microwave process where nearly 100% of the output was composed of recycled Asphalt-Rubber. This project demonstrated that Asphalt-Rubber can be recycled using either microwave technology or conventional mix design technology. Air sampling during paving and recycling determined that employee exposures to air contaminates were well below the Occupational Safety and Health Administration permissible exposure limits, and in most cases below the detection limits.

4. WET PROCESS-NO-AGITATION

No-Agitation Recycled Tyre Rubber Modified Bitumen (No-Agitation RTR-MB), technology was first used in Florida and Texas in the mid-1980s and till nowadays it seems to have been used only in some other states of the USA. This is a form of the wet process where CRM is blended with hot bitumen at the refinery or at a bitumen storage and distribution terminal and transported to the asphalt plant/job site for use. In fact the main intuition behind this technology is to take advantage of the engineered properties of the RTR by using the CRM as an alternative modifier to produce storage stable modified bitumen. These binders are often labeled as terminal blends, although nowadays they may also be field-blended at the asphalt plant. Furthermore, this terminology could sound restrictive also because some of the terminal-blended RTR-MBs, with or without addition of polymers, still present the issue of phase separation and need agitation by special augers or paddles. For this reason someone labeled the storage stable terminal blends as terminal blend-hybrid [60]. Therefore a preferred description for this type of binders is "wet process-No-Agitation" blends [36], which clearly indicates the ability of this material of not requiring special equipment to keep the CRM particles evenly dispersed in the modified binder even during the storage process.

4.1. Production

In this process, no modifications to the asphalt plants are required because the No-Agitation RTR-MBs are manufactured with specific pressure, temperature, time and agitation requirements similarly to polymer modified bitumen. In the wet process-No-Agitation CRM and bitumen are blended through the continuous blending-reaction systems. No reaction tanks are provided. The bitumen is heated under a controlled environment in a tank to an elevated temperature and high shear stress. The CRM is then introduced into the tank and digested into the bitumen. The characteristic swelling process of the RTR-MB technologies is replaced by the depolymerisation/devulcanisation and optimised dispersion of the rubber into the bitumen by using high processing conditions (temperatures between 200 and 300 °C, shear stresses in the order of few thousands RPM and eventually with pressure higher than 1 atm), resulting in a smooth, homogeneous product. In the past, rubber contents for such blends have generally been ⩽10% by weight of bitumen or total binder, but some products now include even 25% CRM [61].

There are some proprietary and other non-proprietary processes. Some of them also include small percentage of other polymers (e.g. SBS), or other additives, to produce a combined homogeneous material that exhibits excellent storage stability and compatibility with the finished binder formulation. Such binders are typically modified with CRM particles finer than the 300 μm (No. 50) sieve size that can be digested (broken down and melted in) relatively quickly and/or can be kept dispersed by normal circulation within the storage. During this process, the operator takes samples and runs a solubility test to ensure the rubber is completely digested. Most manufacturers used a high shear process to make sure the CRM is completed digested. The solubility of the finished product is generally above 97.5%. The binder is then delivered to the hot mix plant by truck as a finished product with no additional handling or processing. After mixing with aggregate material is shipped to the job and no special equipment is required for paving, or odour/fume control. Moreover, the RTR mix is compacted like regular hot mix asphalt [62].

4.2. Storage stability

The wet process-No-Agitation aims to produce bitumen-Recycled Tyre Rubber blends which do not occur in eventual phase separation of the CRM from the binder during storage or transportation. The mechanism of phase separation is related to the differences in the properties, mainly physical, of the material constituents in the blends. After production of the blend, the presence of non-dissolved CRM particles in bitumen leads to phase separation, particularly when the blends are stored at high temperatures. As a result, the swollen CRM particles are considered to settle down quickly due to initial higher density than the bitumen phase [63]. On the other hand, migration of the CRM particles to the top of the storage container due to a reduced density after swelling has also been reported in some research studies [64] and [65]. These different mechanisms lead to an unstable condition which results in a rubberised blend with varied properties after the storage.

The improvement of the settling properties of the No-Agitation RTR-MBs is often linked to the careful selection of the blend's components usage of high-curing conditions [66] which ensures a high level of solubilisation of the CRM within the bitumen matrix. Some patented procedure states that as long as the CRM is fully digested into the binder (solubility is above 97%) it is possible to store them without phase separation [62]. However, at this stage the effect of the modification is significantly reduced. In fact, CRM is shown to a number of studies to effectively modifying the bitumen when it is still in the swollen state and not completely devulcanised/depolymerised (Figure 10c). In order to maintain a convenient level of modification, recent studies have shown that is possible to improve the storage stability of RTR-MBs also by adding substances to operate a chemical stabilisation of the blends [67], by mixing rubberised bitumen blends in presence of a low percentage of polymers [68], with compatibiliser to activate the

CRM [69], or with sulphur to improve crosslinking [64]. Another study [70] shows that hot storage stability of rubberised bitumens could be improved by blowing oxygen gases through these blends. In addition, pre-treatment of CRM particles with hydrogen peroxide has also resulted in a more stable RTR-MBs [71] and [72]. Furthermore, Attia and Abdelrahman [68] suggest improving the stability of modified binders also by lowering the storage temperatures. In any case, the stability of No-Agitation RTR-MBs during prolonged storage seems to be similar to other blended PMBs and the man role in improving the settling properties of bitumen-RTR blends is covered by an accurate selection of the processing conditions (Lo Presti et al. 2012).

4.3. Properties (Like a PMB)

Although such binders may develop a considerable level of rubber modification, rotational viscosity values rarely approach the minimum threshold of 1500 cPs at 177 °C, that is necessary to significantly increase binder contents above those of conventional asphalt mixes without excessive drain down [36]. Recently in the United States the specifications used for terminal blend (PCCAS, 2008) have utilised the PG grading system, named PG-TR system, similar to the ones used for polymer modified bitumen (Figure 4.33). In California, PG-TR grades are specifically targeted for use in the same applications for which PG-PMB binders are used, including dense-graded mixes for thick structural sections. The terminal blends now meet the ASTM definition for minimum CRM content[73].

Within the No-Agitation RTR-MBs production, two main properties are of concern: first, the development of performance-related properties and, second, binder compatibility or storage stability. The literature indicates that performance-related properties develop early in the process while compatibility requires few hours to stabilise [74]. Few investigators have conducted studies on the rheological properties of these binders. For example, Thodesen et al. [75] evaluated several binders using the multiple creep recovery tests. Among the binders studied were an Asphalt-Rubber binder and a No-Agitation RTR-MB that uses 10% CRM and 1% SBS polymer. The results showed that the analysed RTR-MB exhibited the least creep and the highest percentage recoveries under various loading and temperature ranges. Another example is given by Attia and Abdelrahman [68], which investigating the possibility of producing high-performance terminal blends, suitable for SuperPave applications, found that by accurately regulating the processing conditions it is possible to balance the development of performance-related properties and storage stability. The same researchers proposed a procedure consisting in blending bitumen with 5% of CRM and 2% of SBS at elevated shear stress (30–50 Hz) and with the processing conditions shown in Figure 4.38. They found that the final products have performance and binder stability (increasing shearing speed to 50 Hz) comparable to those of patent or proprietary products. Hence, interaction temperature on binder stability was not evaluated. Moreover the authors suggest enhancing the stability of binder by reducing the storage temperature at the plant.

4.4. Applications

No-Agitation RTR-MBs can be used with rubber contents as low as 5% and as high as 25%, depending on the application and the project's requirements. These products can be used in all paving and maintenance applications as a replacement of the polymer modified bitumens: [60]*:*

- Recycled Tyre Rubber Modified Paving Grade Bitumens.
- Standard PG grades – dense graded mixes.
- PG plus grades – high performance mixes.
- Warm mix standard grades – dense and open/gap graded mixes.
- Warm mix polymer modified grades - dense and open/gap graded mixes.
- Hot applied chip seal binders – neat and polymer modified.
- Recycled Tyre Rubber Modified Cutback Bitumens.
- (MC) Medium Cure Graded.
- (SC) Slow Cure Graded.
- Recycled Tyre Rubber Modified Bitumen emulsions.
- Rapid set chip seal.
- Micro surfacing slurry seals.
- Standard slurry seals.
- Cold in-place recycling.
- Cold Mix.

- Recycled Tyre Rubber Modified Seal Coats.
- Recycled Tyre Rubber Modified Surface Sealer.
- Recycled Tyre Rubber Modified Fog Seals.

5. DISCUSSION AND CONCLUSION

High Viscosity RTR-MBs and No-Agitation RTR-MBs are different products and should not be interchanged (Figure 20). However, both provide superior cracking performance at reduced thickness, when compared to conventional dense graded hot mix asphalt pavement. Only when there are construction issues the products are not be expected to perform in a superior manner.

Figure 20. RTR-MB from Wet process (left) and Wet Process-No-Agitation (right), adapted from [73].

In terms of binder properties, the main differences between these products are the viscosity and their storage stability. Viscosity for No-Agitation RTR-MB can range between 500 and 1000 centipoises at 135 °C, much lower than the viscosity for High Viscosity RTR-MB which is in the range of 1500–5000 centipoises at 177.5 °C (Table 8). With regards to the storage stability, No-Agitation RTR-MB born with the idea of obtaining a RTR modified binder comparable to the normal PMBs, therefore on contrary to High Viscosity RTR-MB, it is usually possible to store it as conventional PMBs. Furthermore, a recent study [76] highlights that the storage temperature of the No-Agitation RTR-MB could be significantly decreased (i.e. 15 °C) if compared to the field blended High Viscosity RTR-MB. Considering also the need of the field blends to be stored in agitated tanks, from this point of view the terminal blends leads to significant energy and money savings.

On the other hand, asphalts obtained by using High Viscosity RTR-MBs have more performance history since this process started over in 1960s and they have been used successfully with many applications. With regards to asphalt mixtures, High Viscosity RTR-MB technology is very successful when used with Open-Graded surface courses, where the high air void content of the mix allows an aggregate coating with a much thicker film (36 μm) of high RTR content modified bitumens (about 20%) which leads to an asphalt mix with significantly high binder content (about 7–9%) and with widely proven reduced oxidation, increased durability and increased resistance to reflective cracking. All these benefits are reduced when High Viscosity RTR-MBs is used for Dense-Graded hot mix projects since the dense gradation cannot adequately accommodate the rubber particle size, film thickness is reduced (9 micron) as well as acceptable binder content (about 5%) and the RTR-MBs needs to be produced with much lower rubber content (about 10%). The use of special equipment is not anymore justified by the significant benefits of a thicker coating, therefore in the case of Dense-Graded asphalt mixes the No-Agitation RTR-MBs are the most suitable. On this regard, they are more likely to compete with polymer modified bitumen rather than High Viscosity RTR-MB.

No-Agitation RTR-MBs have been successfully used for a much wider range of products as for instance chips seal applications, open graded and gap graded mixes and emulsions. Basically, No-Agitation RTR-MBs can be used wherever conventional asphalt mixes or asphalt surface treatments are needed. The lower viscosity of No-Agitation RTR-MB implies the usage of less binder per unit area (5–6% binder content) indicating less performance life than if High-Viscosity RTR-MB is used (8–10% binder content). In fact, the ability to inject more binder in the mix translates to better fatigue and reflective cracking performance.

In conclusion, the main advantages of the usage in asphalt mixture of the wet process-No-Agitation in lieu of the wet process – High Viscosity method at the contractor's plant include:

- No need for costly specialised equipment at the asphalt plant.
- No portable plants required for blending of Crumb Rubber with asphalts.
- No additional holding areas for storing the Crumb Rubber product.
- Easiest for the contractor to incorporate into their traditional manufacturing process.

- Mixing, laying and compaction temperatures are comparable to standard mix applications with PMBs.
- Works with all mix designs; does not require any special adjustments to gradation or mix design parameters.
- Being prepared at the refinery, completely eliminates potential problems with heating and blending of Crumb Rubber and asphalt products and
- Eliminates smoke and particulates from entering the atmosphere.
- The binder can be Performance Graded and emulsified
- The few available analyses on the sustainability of the product show that No-Agitation leads to economical saving and environmental benefits.

The aspects that make the wet process-High Viscosity preferable to the wet process-No-Agitation are:

- Evidence of great long term performance, while the long-term performance of existing projects using wet process-No-Agitation are still under evaluation.
- Some have expressed concern that there is no way to verify the amount of CRM used in No-Agitation RTR-MBs.
- Moisture susceptibility of asphalt mixture containing No-Agitation RTR-MB is still not clear.

5.1. Conclusions

The author believes that the widespread use of the RTR-MBs technologies within the road pavement industry is advisable. In fact the several benefits provided to the asphalt pavement performance, and to the overall sustainability of the infrastructure, are so evident that it is strongly advised to consider RTR-MBs technologies as a first option to the binders currently used in road pavements. Companies, road authorities, etc. have to evaluate if it is convenient to use the High Viscosity wet process technology, which proved widely to provide several benefits, in particular it allows highway designers to reduce pavement layer thickness due to the proven properties of rubberised bitumen, but presents some challenges as: the need for suitable blending and mixing equipment, the cost of such equipment and the degree of difficulty in preparing asphalt mix design. The other option is to choose the wet process-No-Agitation technology which solves several issues but leads to asphalts pavements with, so far, uncertain long term performance. In both cases the implementation of these technologies still presents issues as: the lack of availability of suitable RTR processing facilities in the vicinity and the cost of such facilities, the lack of rubberised bitumen binder and mix standards, the uncertainty on the environmental performance of these products and the lack of trained personnel particularly for accurate laboratory analyses, of raw materials and final blends, which are necessary to optimise the modification process and the performance of the final product. Education and training are indeed another key aspect for a successful application of RTR-MBs in road asphalt mixtures also because new procedures are being developed to perform an enhanced laboratory design of these technologies [77] as well as reducing limitations to their applicability in the field. Examples are given by the increasing number of rubberised asphalt applications with additives able to reduce paving temperatures (i.e. warm mix technologies) [21] and [57], but also from the invention of a new form of product delivery as the case of the rubberised pellets [78]. Another important aspect to clarify is the cost of this technology: it is true that initial costs are still an issue but the rapid cost increase of bitumen and results of lifecycle cost analyses indicate that RTR-MBs are in many cases an economically convenient option. At last, there is evidence that a closer involvement of local and national governments with policies supporting the RTR industries and the major investments on training and research could definitely help to decrease the costs of implementing RTR-MB technologies as well as solving the several issues indicated within this manuscript. A proof of this statement is given by the state of California (USA) where since 2005 a government mandate [79] supports the increasing adoption of these technologies in asphalt mixture and from the January 2013 the usage of rubberised asphalt pavements needs to reach at level of at least 35% of the total weight of asphalt paving materials produced in the country. The main reason of this mandate is the prediction of substantial savings in the long term.

REFERENCES

1. ETRMA statistics. ; 2012.

2. ETRMA ELTs. End-of-Life Tyres. ; January 2011.

3. Rahman MM. Characterisation of dry process crumb rubber modified asphalt mixtures. Thesis submitted to the University of Nottingham for the degree of Doctor of Philosophy. s.l.: University of Nottingham, School of Civil Engineering; 2004.

4. Miskolczi N, Nagy R, Bartha L, Halmos P, Fazekas B. Application of energy dispersive X-ray fluorescence spectrometry as multielemental analysis to determine the elemental composition of crumb rubber samples. Microchem J 2008;88(1):14–20.

5. Diffusion kinetics of bitumen into waste tyre rubber. Artmandi, I and Khalid, H A. Savannah: s.n., 2006. Journal of the Association of Asphalt Paving Technologists. Proceedings of the Technical Sessions. vol. 75. Georgia, USA; 2006. p. 133–64.

6. Peralta E. Study of the interaction between bitumen and rubber – PhD thesis. ; 2009 [Cited: 08.03.13].

7. Checkthatcar.com. ; 2013.

8. Rimondi G.; 2009.

9. Walker D. Understanding how tires are used in asphalt. ; 2013.

10. Oliver JWH. Modification of paving asphalts by digestion with scrap rubber. [ed.] Transportation Research Board. Transportation Research Record 821; 1981.

11. Roberts FL, Kandhal PS, Brown ER, Dunning RL. Investigation and evaluation of ground tyre rubber in hot-mix asphalt. Alabama, USA: NCAT Auburn University; 1989.

12. Memon N. Characterisation of conventional and chemically dispersed crumb rubber modified bitumen and mixtures. University of Nottingham. Nottingham, UK: s.n. PhD thesis; 2011.

13. Heitzman M. Design and construction of asphalt paving materials with Crumb Rubber Modifier. Transportation Research Record 1339; 1992.

14. Epps JA. Uses of recycled rubber tyres in highways. Washington, DC: Synthesis of Highway Practice No.198, TRB National Research Council. NCHRP Report; 1994.

15. Kuennen T. Asphalt rubber makes a quiet comeback. Better Roads Magazine; May 2004.

16. Terrel RL, Walter JL. Modified asphalt materials – the European Experience. In: s.l.: AAPT proceedings. vol. 55. 1986. p. 482–518.

17. Bitumen–rubber: lessons learned in South Africa. Visser AT, Verhaege B. Portugal: ISBN: 972-95240-9-2, 2000. Asphalt Rubber; 2000.

18. Widyatmoko I, Elliot R. A review of the use of crumb rubber modified asphalt worldwide. UK : Waste & Resources Action Programme (WRAP); 2007.

19. Souza R. Experiences with use of reclaimed rubber in asphalt within Europe. Birmingham: s.n. Rubber in Roads; 2005.

20. Mavridou S, Oikonomou N, Kalofotias A. Worldwide survey on best (and worse) practices concerning rubberised asphalt mixtures implementation (number of different cases, extent of application. Thessaloniki: EU-LIFE+ Environment Policy and Governance. ROADTIRE, D2.1.1; 2010.

21. Hicks G, Cheng D, Teesdale T, Assessment of Warm Mix technologies for use with Asphalt Rubber paving application. Tech-Report-103TM; 2010.

22. Antunes M, Baptista F, Eusébio MI, Costa MS, Valverde Miranda C. Characterisation of asphalt rubber mixtures for pavement rehabilitation projects in Portugal. Asphatl Rubber 2000; 2000.

23. Gallego J, Del Val MA, Tomas R. Spanish experience with asphalt pavements modified with tire rubber. Asphalt Rubber 2000; 2000.

24. Santagata FA, Canestrari F, Pasquini E, Mechanical characterisation of asphalt rubber – wet process. Palermo: s.n. In: 4th International SIIV Congress; 2007.

25. Dasek O, Kudrna J, Kachtík J, Spies K. Asphalt rubber in Czech Republic. Munich: s.n. Asphatl rubber 2012; 2012.

26. Nordgren T, Tykesson A, dense graded asphalt rubber in cold climate conditions. Munich: s.n. Asphalt Rubber 2012; 2012.

27. Pinto A, Sousa J, The first brazilian experience with in situ field blend rubber asphalt. Munich: s.n. Asphalt Rubber 2012; 2012.

28. Nourelhuda M, Ali G. Asphalt-rubber pavement construction and performance: the sudan experience. Munich: s.n. Asphalt Rubber 2012; 2012.

29. Bahia H, Davis R, Effect of Crumb Rubber Modifiers (CRMs) on performance related properties of asphalt binders. AAPT 1994; 1994.

30. Cheovits JG, Dunning RL, Morris GR. Characteristics of asphalt-rubber by the slide plate microviscometer. Association of Asphalt Paving Technologists. vol. 51. 1982. p. 240–61.

31. Review of Utilization of Waste Tires in Asphalt. Shatanawi K, Thoedsen C. s.n. In: Globhal Plasti Environmental Conference. Orland, FL, USA; 2008.

32. Abdelrahman MA, Carpenter SH. The mechanism of the interaction of asphalt cement with crumb rubber modifier (CRM). Transport Res Rec 1999;1661: 106–13.

33. Lo Presti D, Airey G, Partal P. Manufacturing terminal and field bitumen-tyre rubber blends: the importance of processing conditions. Proc-Soc Behav Sci 2012;53:485–94.

34. DGRoadMachinery.com. www.dgroadmachinery.com.; 2013.

35. WRIGHTasphalt. Terminal blendedterminal blendedtire rubber modified tire asphalt cementasphalt cement.

36. Caltrans. Asphalt Rubber Usage Guide. s.l.: State of California Department of Transportation, Materials Engineering and Testing Services; 2006.

37. ASTM D6114/D6114M. s.l.: American Society for Testing and Materials; 2009. Standard Specification for Asphalt-Rubber Binder.

38. AsAc. The use of Modified Bituminous Binders in Road Construction. Technical Guideline of the South African Asphalt Acadamey TG1; 2007.

39. Potgieter CJ, Bitumen Rubber Asphalt in South Africa Conventional Techniques. Vienna: s.n. In: 3rd Eurasphalt & Eurobitume Congress; 2004.

40. SABITA. manual 19. Guidelines for the design, manufacture and construction of bitumen rubber asphalt wearing courses; 2009.

41. Austroads. Specification framework for polymer modified binders and multigrade bitumens". Austroads Technical Report AP-T41/06; 2006.

42. OGFC Meets CRM. Where the Rubber meets the Rubber. 12 Years of Durable Success. Way, G.B. Vilamoura: s.n. In: Asphalt Rubber 2000 Conference; 2000.

43. Santagata FA, Antunes I, Canestrari F, Pasquini E. Asphalt Rubber: Primeiros Resultados em Itália. Estoril: s.n. Estrada 2008, V Congresso Rodoviário Português; 2008.

44. Sousa JB, Experiences with use of reclaimed rubber in asphalt within Europe. Birmingham: s.n. Rubber in Roads; 2005.

45. Bertollo SM, Mechanical properties of asphalt mixtures using recycled tyre rubber produced in Brazil – a laboratory evaluation. Washington: s.n. In: 83rd TRB Annual Meeting; 2004.

46. Kaloush KE, Witczak MW, Sotil AC, Way GB, Laboratory evaluation of asphalt rubber mixtures using the dynamic modulus (E/) Test. Washington: s.n. In: 82nd TRB Annual Meeting; 2003.

47. Jung J, Kaloush KE, Way GB, Life cycle cost analysis: conventional versus asphalt rubber pavements. s.l.: Rubber Pavement Association., 2002; 2003.

48. Noise Reducing Asphalt Pavements: a Canadian Case Study. Leung F et al. s.n. In: 10th International conference on asphalt pavements. Quebec City; 2006.

49. The Successful World Wide Use of Asphalt Rubber. Antunes I et al. Cosenza: s.n. In: 16th Convegno Nazionale SIIV; 2006.

50. Pasquini E. Advanced characterisation of innovative environmentally friendly bituminous mixtures. s.l.: Università Politecnica delle Marche. PhD thesis; 2009.

51. Antunes I, Asphalt rubber: Il Bitume Modificato con Polverino di Gomma di Pneumatico Riciclata. L'Aquila s.n. Varirei – V International Congress of Valorisation and Recycling of Industrial Waste; 2005.

52. Advances in technology of asphalt paving materials containing used tyre rubber. Takallou HB and Sainton A. Transportation Research Reccords. vol. 1339. 1992. p. 23–29.

53. Hicks R, Epps JA. Life cycle costs analysis of asphalt rubber paving materials. Tempe, AZ: The Rubber Pavements Association; 2003.

54. Carlson D. Developing trends in rubberized asphalt. US EPA SMM Web Academy Webinar Series, www.epa.gov.; 2013.

55. Crockford WW, Recycling crumb rubber modified asphalt pavements. s.l.: Texas Transportation Institute. Report FHWA/TX-95/1333-1F; 1995.

56. Khattak S, Syme B. Terminal blend tyre rubber asphalt technical information. s.l.: City of Colorado Springs and Nolte Associates, Inc.; 2010.

57. Jones D, Wu R, Barros C, Peterson J. Research findings on the use of rubberised warm-mix asphalt in California. Asphalt-Rubber 2012; 2012.

58. Carlson D, Zhu H, An anchor to crumb rubber markets. ; 1999.

59. Youssef Z, Hovasapian PK. Olympic boulevard asphalt rubber recycling project. s.l.: City of Los Angeles Department of Public Works; 1995.

60. Quire D. Tire rubber modified asphalt, asphalt pavement conference presentations. ; March 2012.

61. Raetexindustries.com, Terminal_Blend_vs_AR.pdf. .

62. Asphalt Institute online magazine, Terminal Blended Rubberised Asphalt Goes Mainstream – Now PG Graded. ; 2008.

63. Navarro FJ, Partal P, Martinez-Boza FJ, Gallegos F. Thermo-rheological behaviour and storage stability of ground tyre rubber-modified bitumens. Fuel 14–15 October 2004;83.

64. Ghaly NF. Effect of sulfur on the storage stability of tire rubber modified asphalt. World J Chem 2008;3(2):42–50.

65. Perez-Lepe A, Martinez-Boza FJ, Gallegos F. High temperature stability of different plymer-modified bitumens: a rheological evaluation C. J Appl Polym Sci 2007.

66. Glover CJ, Davison RR, Bullin JA, Estakhri CK, Williamson SA, Billiter TC, Chipps JF, Chun JS, Juristyarini P, Leicht SE, Wattanachai P. A comprehensive laboratory and field study of high-cure crumb-rubber modified asphalt 21 materials. USA: Texas Transportation Institute; 2000. Report 1460-1.

67. Biro S, Bartha L, Deak G, Geiger A, Pannonn E. Chemically stabilized asphalt rubber compositions and a mechanochemical method for preparing the same. HU226481, WO/2007/068990 – 2007, EP1960472 – 2008; 2008.

68. Attia M, Abdelrahman MA. Enhancing the performance of crumb rubbermodified binders through varying 1 the interaction conditions. M. 6. Int J Pavement Eng 2009;10:423–34.

69. Cheng G, Shen B, Zhang J. A study on the performance and storage stability of crumb rubber-modified 11 asphalts. Petrol Sci Technol 2010.

70. William GE, Polymer enhanced asphalt. WO patent, 97455488; 1997.

71. Maldonado PJ, Phung TK, Process for preparing polymers-bitumen compositions. US patent, 4145322; 1979.

72. Planch JP. Method for the preparation of bitumen polymer compositions. WO Patent, 9002776; 1990.

73. Shatnawi S, Comparisons of runnerized asphalt binders: asphalt-rubber and terminal blend. Munich: s.n. In: Asphalt Rubber proceedings 2012; 2012.

74. Abdelrahman MA, Carpenter SH. The mechanism of the interaction of asphalt cement with crumb rubber modifier (CRM). USA : s.n.. Trans Res Rec 1999;1661:106–13.

75. Thodesen C, Biro S, Kay J. Evaluation of current modified asphalt binders using the multiple stress creep recovery test. s.n. In: Proceedings of AR2009 conference. Nanjing, China; 2009.

76. Wu C, Liu K, Tang J, Li A. Research on the terminal blend rubberized asphalt with high-volume of rubber crumbs and its gap graded mixture. Munich: s.n. Asphalt rubber 2012; 2012.

77. Celauro B, Celauro C, Lo Presti D, Bevilacqua A. Definition of a laboratory optimization protocol for road bitumen improved with recycled tire rubber. s.l. Constr Build Mater 2012;37:562–72.

78. Amirkhanian S, Kelly S. Development of polymerized asphalt rubber pelleted binder for HMA mixtures. Munich: s.n. In: Asphalt rubber proceedings 2012; 2012.

79. Leginfo.ca.gov.; 2013.

10
CHAPTER

Assessment of Concrete Produced with Foundry Waste as Partial Replacement for River Sand

T. C. Nwofor [1], C. Ukpaka [2]

[1]Department of civil Engineering, University of Port Harcourt, Port Harcourt, Nigeria

[2]Jospaka Ventures Nigeria Limited, Port Harcourt, Nigeria

ABSTRACT

The assessment in this research is targeted at determining the percentage of foundry waste suitable to be used as a partial replacement for fine sand aggregate. Experimental result reveals that 15% of foundry sand is appropriate as partial replacement of fine sand aggregate for preparation of grade 20 concrete by application of compressive strength equation as targeted strength of concrete were established at 28 days. The mathematical tools known as least square model was also applied to simulate the best equation of the curve for the purpose of prediction of compressive strength for the design of WFS concrete. The equation of best fit is given as 6 = 13.77wfs + 13.65 for 7 days and 28 days and 6 = 1.303wfs + 17.49 for 21 days. The developed equation illustrates the usefulness of least square model in monitoring the compressive strength of concrete upon the variation of percentage replacement with waste foundry sand.

Keywords: Concrete, Waste Foundry Sand, Fine Sand and Compressive Strength

1. INTRODUCTION

The evaluation of engineering aggregate materials as good alternative replacement or partial replacement of aggregate in concrete production that will be affordable and durable is of global interest. Aggregates are the significant constituents in concrete. They give weight to the concrete and reduce shrinkage [1-3].

Assessment of foundry sand as a replacement for fine aggregate on fresh and hardened characteristics of concrete revealed that compressive strength of concrete decreases with increase in foundry sand [4-5]. Research conducted on fresh and hardened properties of grade 20 concrete, by substituting waste ferrous and non-ferrous foundry sand revealed that the compressive strength increased normally at 7days, however the strength gradually decrease as waste foundry sand content increased at 21 and 28 days, which is similar for other kind of replacements [6-14]. In terms of flexural strength, concrete strength increased with the increased in founding sand content at 28 days.

This research is centered at investigating the foundry sand as an alternative for partial replacement of fine sand aggregate in the production of concrete, thereby making the waste founding sand in Nigeria a useful waste by applying the empirical formula for compressive strength for adequate mix design.

2. MATERIALS AND METHOD

2.1. Cement

R Dangote 3x Portland cement were used for batching and mixing of all specimens.

2.2. Standard Grading Curve

The grading patterns of aggregate can be shown on tables or charts. Expressing grading limits by means of charts gives a good pictorial view. The correlation of grading pattern of a number of samples can be made at one glance. This is the reason, often grading of aggregates is shown by means of grading curves. Practical knowledge has revealed that in practice it is difficult to get the aggregate to conform to any one particular standard curve exactly where concrete of high strength and good durability is required, time aggregate zones may be used, but the concrete mix should be properly designed. If the fine aggregate grading becomes progressively finer, that is from aggregate to coarse aggregate should be progressively reduced. The most suitable fine to coarse rating of aggregate to be applied for any particular mix will depend on the grading particle shape and surface texture of both fine and coarse aggregates.

2.3. Water

The quantity and quality of water used during the batching of the concrete is of good one and based on water / cement adopted as water is a significant ingredient of concrete as it actively participates in the chemical reaction with cement.

2.4. Moulds

The moulds used were cleaned and oiled properly in order to remove the concrete from the mould easily after casting is done. The cube moulds size is 150 x 150mm and are used for the preparation of the concrete specimens in accordance with BS: 12390 part 1.

2.5. Mix Design of Concrete

In this research selecting suitable ingredients of concrete and determining their relative proportions with the object of producing concrete of certain minimum strength and durability as economically as possible is based on specification of BS:1881: part 125 – 1986. Hence, 1:2:4 design mix was used, where the river fine sand content were partially replaced with foundry sand aggregate at a constant water-cement of 0.5.

Three number of concrete cubes were cast for control mix (0% Foundry sand, tagged M-0) and design mixes at each replacement level. Fine aggregate were partially replaced with foundry sand at 5,10,15,20 and 25% replacement and specimens tagged M-1, M-2, M-3, M-4 and M-5 respectively.

3. RESULTS AND DISCUSSION

3.1. Aggregate Grading Curve

River Sand (Fine Aggregate)

Sieve analysis investigation on fine aggregate was carried out using BS 410-2000. The analysis reveals that the fine aggregate conforms to the grading pattern of zone 3 as illustrated Figure 1. The bulk density, specific gravity and the percentage passing of the fine sand aggregate is presented in Table 1 and 2.

Table 1. Physical Properties of Fine Aggregate

S/N	Properties	Observed values
1	Bulk density (kg/m^3)	1639
2	Specific gravity	2.68

Table 2. Sieve Analysis of Fine Aggregate

Weight of sample taken = 500gm

Particle Description		Sieve Mm	Total Wt. (g)	Wt. Retained	% Retained	% Passing	Specification
	Cobbles	100.00					
		90.00					
		75.00					
		63.00					
	Coarse	50.00					
		31.80					
GRAVEL		25.00					
		19.00					
		12.50					
	Fine	10.00	0.0	0.0	0.00	100.00	
		5.00	14.00	14.0	2.80	97.20	**90-100**
		2.36	30.00	16.0	3.20	94.00	**85-100**
	Medium	1.18	92.00	62.0	12.40	81.60	**75-100**
		0.60	166.00	74.0	14.80	66.80	**60-79**
SAND		0.300	394.00	228.0	45.60	21.20	**12- 40**
	Fine	0.150	478.00	84.0	16.80	4.40	**0-10**
		0.075					
FINES	Clay or Silt	<0.075					

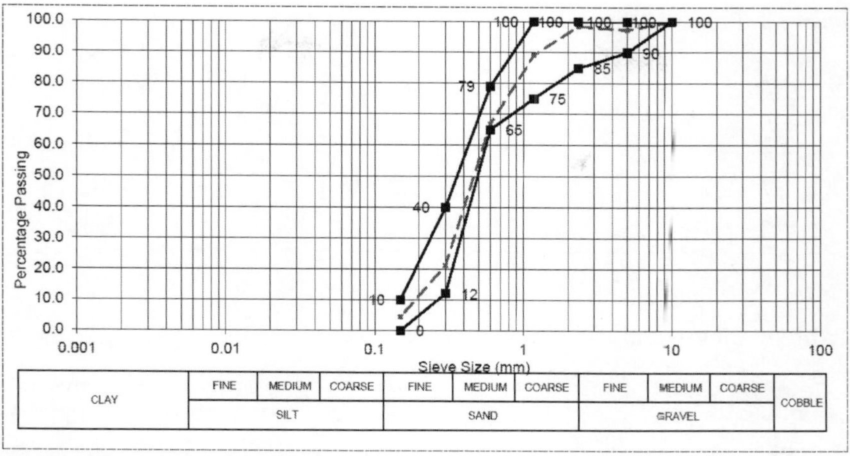

Figure 1. Particle size distribution curve of fine aggregate

Coarse Aggregate

The available coarse aggregate in Nigeria having the maximum size of 20mm were used in this research investigation for the preparation of the concrete specimens. Figure 2 shows the pattern of the particle size distribution of the coarse aggregate. The Maximum size, bulk density, specific gravity and percentage passing are presented in Table 3 and 4 as shown below.

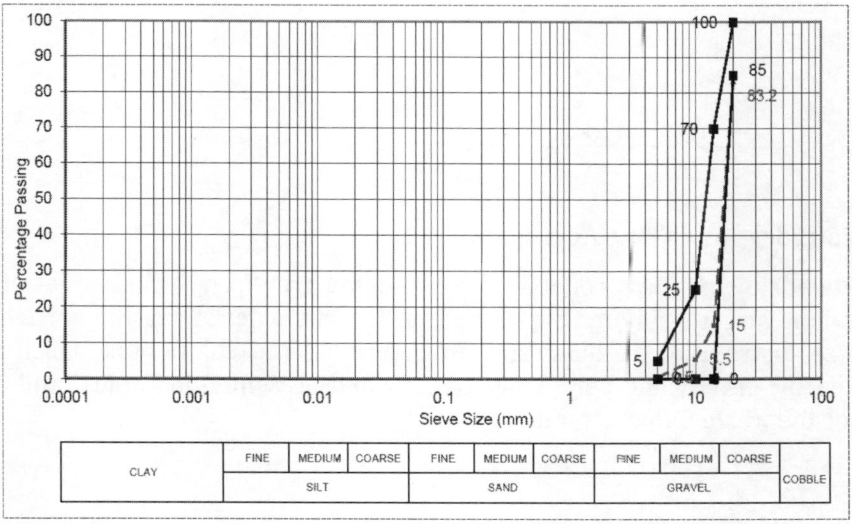

Figure 2. Particle Size Distribution Curve of Coarse Aggregate

Table 3. Physical Properties of Coarse Aggregate

S/N	Properties	Values
1	Maximum size	20mm
2	Bulk density (kg/m^3)	1850
3	Specific gravity	2.74

Table 4. Sieve Analysis of Coarse Aggregate

Particle Description		Sieve Mm	Total Wt. (g)	Wt. Retained	% Retained	% Passing	Specification Grading limit
	Cobbles	100.00					
		90.00					
GRAVEL	Coarse	75.00					
		63.00					
		50.00					
		31.80					
		25.00				100.00	100
	Fine	20.00	168.00	168.0	16.80	83.20	85-100
		14.00	850.00	682.0	68.20	15.00	0-70
		10.00	945.00	95.0	9.50	5.50	0-25
		5.00	995.00	50.0	5.00	0.50	0-5
SAND	Medium						
	Fine						
FINES	Clay or Silt						

Foundry Sand (Fine Aggregate)

Foundry waste was collected from stemma gray-iron, Asaba-street off Afikpo street in Ebonyi State, Nigeria. The characteristic of the particle size distribution is illustrated in Figure 3. The color, bulk density, specific gravity and percentage passing and presented in Table 5 and 6 for the waste foundry sand.

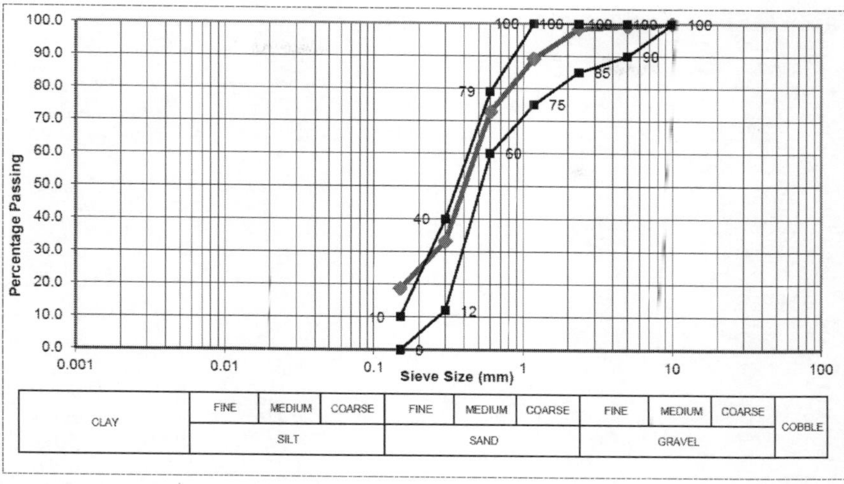

Figure 3. Particle Size Distribution Curve of Foundry Sand

Table 5. Physical Properties of Foundry Sand

S/N	Properties	Values
1	Color	Gray (blackish)
2	Bulk density (kg/m³)	1598
3	Specific gravity	2.61

3.2. Compressive Strength

The evaluation of the compressive strength of the concrete made containing waste foundry sand at different replacement levels (0%, 5%, 10%, 15%, 20% and 25%) at the end of different curing period (7days, 21days, and 28days) as presented in Table 7. The results were obtained by application of compressive strength equation, given as

$$\sigma = \frac{P}{A} \quad (1)$$

Table 6. Sieve Analysis of Foundry Sand

Weight of sample taken = 500gm

Particle Description		Sieve mm	Total Wt. (g)	Wt. Retained	% Retained	% Passing	Specification
GRAVEL	Cobbles	100.00					
		90.00					
	Coarse	75.00					
		63.00					
		50.00					
		31.80					
		25.00					
		19.00					
	Fine	12.50					
		10.00	0.0	0.0	0.00	100.00	
SAND		4.75	4.00	4.0	0.80	99.20	**90-100**
		2.36	30.00	26.0	5.20	94.00	**85-100**
	Medium	1.18	72.00	42.0	8.40	85.60	**75-100**
		0.60	136.00	64.0	12.80	72.80	**60-79**
		0.300	334.00	198.0	39.60	33.20	**12- 40**
	Fine	0.150	406.50	72.5	14.50	18.70	**0-10**
		0.075		56.0	11.20	11.62	
FINES	Clay or Silt	<0.075			0.00	7.29	

Where; 6 = compressive strength (N/mm^2), P = maximum load (N), A = cross-sectional area of cube (mm^2).

Figure 4 illustrate the relationship between compressive strength (6) of concrete and waste foundry sand for various partial replacement of fine sand aggregate with waste foundry sand (WFS) proportion. The linear regression model for 7 days reveals that the compressive strength relates to the WFS by the equation, 6 = 13.77WFS + 13.65 and coefficient of correlation R = 0.70. Similarly both parameters are related by the same equation at the 28th day while the 21th day was related by, 6 = 1.303WFS + 17.49 with R = 0.14.

Also from Figure 7, it was seen that the concrete produced with varying percentages of foundry sand showed higher compressive strength than the control specimen (0% WFS) up to a 15% replacement level after which there was a gradual decrease in strength from the 20 to 30% replacement level.

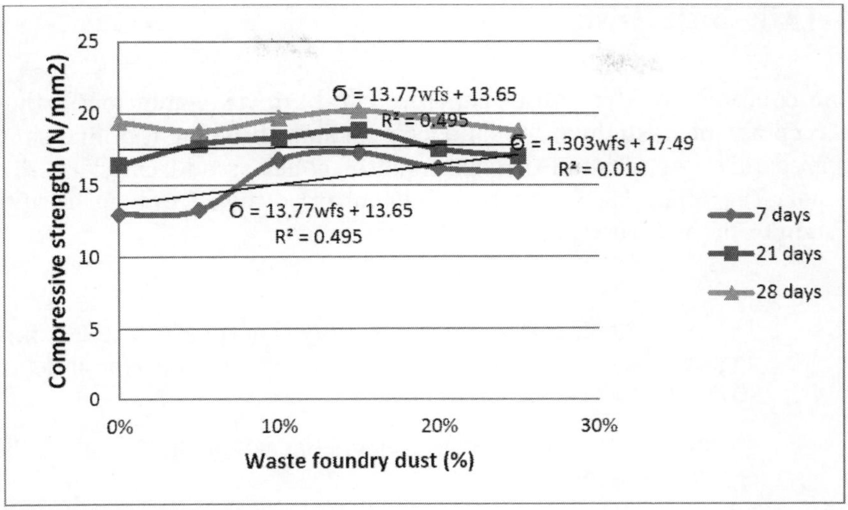

Figure 4. Plot of Compressive Strength against Foundry Sand Replacement Level

Table 7. Compressive Strength (N/mm²) of Concrete at Different foundry sand Replacement Level

Foundry sand content (%)	Designation	Compressive strength, N/mm²		
		7 days	21 days	28 days
0%	M-0	12.95	16.41	19.38
5%	M-1	13.24	17.87	18.76
10%	M-2	16.76	18.31	19.64
15%	M-3	17.24	18.84	20.18
20%	M-4	16.18	17.51	19.47
25%	M-5	15.91	16.98	18.76

4. CONCLUSIONS

The compressive strength of concrete can be investigated for further percentage of waste foundry sand as a partial replacement of fine sand aggregate by application of least square model as addressed by this paper. The following specific conclusion are drawn from the results obtain from the investigation;

1. The use of foundry sand as partial replacement of fine aggregates in making concrete increases the compressive strength at all replacement levels.
2. A maximum compressive strength is achieved at 15% replacement level.

REFERENCES

1. Shetty M.S. (2005). Concrete technology theory and practice. First Illustrative Revised Edition.
2. Murdock L.J. (1960). The workability of concrete, Magazine of Concrete Research.
3. Jagus P.J. (1958). Alkali aggregate reaction in concrete construction. Road Research Paper No. 11.
4. Khatib, J.M & Ellis, D.J. (2001). Mechanical properties of concrete containing foundry sand. ACI. Special Publication; (SP-200): 733-748.
5. Khatib, J.M., Herki B.A. & Kenai, S. (2013). Capillarity of concrete incorporating waste foundry sand. Construction and Building Material 47, 867-871.
6. Eknath, P.S. & Desai, D.B. (2009). Application of foundry waste sand in manufacture of concrete. IOSR journal of Mechanical and Civil Engineering, 43-48.
7. Gurpreet S. & Rafat S. (2012). Effect of waste foundry sand (WFS) as partial replacement of sand on strength. Journal of Construction and Building Material 26, 416-422.

8. Nwofor, T.C. & Eme D.B. (2010). Stability of concrete work with partial substitute of rice husk ash (RHA) for cement in Nigeria. Journal of Chemical, Mechanical and Engineering Practise, 1(2&3), 42-47.

9. Sule S. & Nwofor, T.C. (2011). Analytical models for the prediction of mechanical properties of rice husk ash concrete. International Journal of Current Research, 3(11), 368-370.

10. Nwofor, T.C. & Sule, S. (2012). Stability of groundnut shell (GSA)/ordinary Portland cement (OPC) concrete in Nigeria. Advances in Applied Science Research, 3(4), 2283-2287.

11. Nwofor, T.C. & Sule, S. (2012). Investigating geometric characteristics of cement concrete materials. International Journal of Innovative Research in Advance Engineering, 1(9), 74-82.

12. Nwofor, T.C., Sule, S. & Eme, D.B. (2015). A comparative study of the methods of concrete mix design using crushed and uncrushed coarse aggregates. International Journal of Scientific and Engineering Research, 6(8), 1182-1194.

13. Eme, D.B., Nwofor, T.C. & Umukoro, E.S. (2015). Stability of asphalt cement concrete produced from waste plastics as replacement for aggregate. International Journal of Civil Engineering and Technology, 6(5), 65-75.

14. Eme, D.B. & Nwofor, T.C. (2015). Investigating the Marshall Stability requirements of asphalt concrete mix with ground scrap tyres as aggregate. International Journal of Civil Engineering and Technology, 6(9), 1-7.

Index